# Accelerating Democracy

# Accelerating Democracy

TRANSFORMING GOVERNANCE
THROUGH TECHNOLOGY

*John O. McGinnis*

PRINCETON UNIVERSITY PRESS

Princeton & Oxford

Copyright © 2013 by Princeton University Press
Published by Princeton University Press, 41 William Street, Princeton, New Jersey 08540
In the United Kingdom: Princeton University Press, 6 Oxford Street,
Woodstock, Oxfordshire OX20 1TW

press.princeton.edu

Library of Congress Cataloging-in-Publication Data

McGinnis, John O., 1957-
Accelerating democracy : transforming governance through technology /
John O. McGinnis.
     pages cm
Includes bibliographical references and index.
ISBN 978-0-691-15102-1
1. Information technology—Political aspects. 2. Technological innovations-
Political aspects. 3. Democracy. 4. Democratization. I. Title.
HC79.I55M3665 2013
303.48'3—dc23
                                   2012033068

British Library Cataloging-in-Publication Data is available

This book has been composed in Sabon

Printed on acid-free paper. ∞

Printed in the United States of America

10 9 8 7 6 5 4 3 2 1

For Ardith

# Contents

INTRODUCTION     1

CHAPTER ONE
The Ever Expanding Domain of Computation     9

CHAPTER TWO
Democracy, Consequences, and Social Knowledge     25

CHAPTER THREE
Experimenting with Democracy     40

CHAPTER FOUR
Unleashing Prediction Markets     60

CHAPTER FIVE
Distributing Information through Dispersed Media
and Campaigns     77

CHAPTER SIX
Accelerating AI     94

CHAPTER SEVEN
Regulation in an Age of Technological Acceleration     109

CHAPTER EIGHT
Bias and Democracy     121

CHAPTER NINE
De-biasing Democracy     138

CONCLUSION
The Past and Future of Information Politics     149

*Acknowledgments*     161

*Appendix*     163

*Notes*     165

*Index*     203

# Accelerating Democracy

MOST OF US are caught up in the quickening whirl of technological change. As consumers we can readily recognize the benefits created by the quicker technological tempo—ever smarter phones, more effective medicines, and faster connections to those around us. We thrive as companies leapfrog one another to create the next new thing. We rarely pause, however, to consider what such technological progression means for our lives as citizens.

Yet the central political problem of our time is how to adapt our venerable democracy to the acceleration of the information age. Modern technology creates a supply of new tools for improved governance, but it also creates an urgent demand for putting these tools to use. We need better policies to obtain the benefits of innovation as quickly as possible and to manage the social problems that speedier innovation will inevitably create—from pollution to weapons of mass destruction.

Exponential growth in computation is driving today's social change. The key advantage for politics is that increases in computational power dramatically improve information technology. Thus, unlike most technological innovations of the past, many innovations today directly supply new mechanisms for evaluating the consequences of social policies. Our task is to place politics progressively within the domain of information technology—to use its new or enhanced tools, such as empiricism, information markets, dispersed media, and artificial intelligence, to reinvent governance.

For instance, the Internet greatly facilitates betting pools called information or prediction markets that permit people to bet on the occurrence of future events. Such markets already gauge election results more accurately than polls do. If legalized and modestly subsidized, they could also foretell many policy results better than politicians or experts alone. We could then better predict the consequences of changes in educational policy on educational outcomes or a stimulus program on economic growth. In short, prediction markets would provide a visible hand to help guide policy choices.

The Internet today also encourages dispersed media like blogs to intensify confrontations about contending policy claims. Previously a less diverse mainstream media tended to settle for received wisdom. Our more competitive media culture permits the rapid recombination of innovative policy proposals and expert analysis no less than our more competitive scientific culture provides an incubator for new computer applications. A novel plan for reducing unemployment is immediately analyzed, critiqued, and compared to other programs.

Because of this greater computational capacity, society can also use more effective methods of social science to evaluate empirically the results of policies. Like prediction markets and dispersed media, the turn to empiricism benefits from competitive structures. Different jurisdictions, from states to school districts, try to gain advantages over one another by adopting better policies. With our more sophisticated empirical tools, we can then assess the effect of their distinctive policies, gauging the degree to which gun control helps prevent crime or whether longer school hours improve student learning.

Thus, the technological transformation of society contains within itself the dynamo of its own management, but only if we create laws and regulations to permit the information revolution to wash through our democratic structures. For instance, Congress must legalize online prediction markets and systematically encourage policy experimentation within the framework of its legislation. The Supreme Court must assure that the changing media can deliver information to citizens, particularly at election time. We cannot tolerate social learning that moves at glacial speed when technological change gallops apace; we cannot put up with a government that inaccurately assesses policy results with outdated methods when new smarter mechanisms are within its reach. The technological revolution is giving us progressively better hardware to gather the information in the world to improve policy outcomes. But government structures and rules provide the essential social software to make that hardware work effectively on behalf of society.

With the advent of new technology, the ideal structure for social governance today starkly contrasts with previous visions of modern government, like that celebrated in the New Deal, which relied on centralized planning. There the focus was also on improved governance through the use of social information, but the analysis was to be handed down from the top—from experts and bureaucrats. Today technology permits knowledge to bubble up from more dispersed sources that are filtered through more competitive mechanisms, sustaining a more decentralized and accurate system of social discovery. We can acquire general expertise without being beholden to particular experts. The nation can retain and improve the best of the model of governance we have—a politics that seeks to be informed by expertise and social-scientific knowledge—while shedding the error-prone arrogance and insularity of a technocracy.

The promise of modern information technology for improving social governance should not be confused with an enthusiasm for using technology simply to increase democratic participation. Often labeled "digital democracy," this perspective animates President Barack Obama's promise to respond to Web-based petitions that collect five thousand signatures.[1] But more equal participation is not sufficient to assess more accurately the consequences of social policy, because citizens do not possess equal knowledge.

Modern information technology instead allows us to root improved governance in a realistic assessment of human nature. It permits competitive mechanisms and the scientific method to harness man's self-interest and unequally distributed knowledge for the public good. More utopian visions of social reform, which rely solely either on the opinions of an elite or the unrefined sentiment of the people to remake society, are worse than political blunders. They are anachronisms in the information age, with its more accurate methods for sifting information and translating it into the knowledge needed to evaluate policy.

The use of market mechanisms and the scientific method also could lower the political temperature. Such tools encourage a greater recognition of the uncertain effects of all human action and thereby bring a greater dispassion to the business of social reform. This style of politics makes it more likely that society will act on the best evidence available to make steady improvements and avoid the worst catastrophes.

Contemporary technology not only supplies tools for better decision making but also creates a demand for their deployment to both speed the benefits of innovation and handle its dangers. Technological innovations today hold the potential to provide a great boon to humanity. Advances in biotechnology and other fields promise longer life through medical innovation, and those in computation generate greater wealth through enhanced productivity. But the pace of these beneficial inventions depends in part on government decisions about taxes, government investments in basic science, and laws on intellectual property. Government can create a virtuous circle by using technology to sustain social processes that in turn create a faster cycle of valuable technologies. In contrast, bad government policy on innovation is today more costly than ever, because it can squander unparalleled opportunities for technological advance.

Even more important, better government is needed to address the downsides of faster and likely accelerating technological change. Energy-intensive machines began the process of injecting greenhouse gases into the atmosphere at the time of industrialization. Yet almost no one recognized this development until relatively recently, in large part because we did not have the necessary predictive tools. Thus our prior failure to foretell global warming signals the need for earlier and more accurate assessment of the possibly dangerous by-products of more advanced and more rapidly developing technology.

Domestically, fast-moving technology can be socially disruptive. Improving machine intelligence can at times complement the value of an employee, enhancing his or her productivity. But it can also at times provide a complete substitute for human labor. As computer search capabilities become more effective, the human premium on simply knowing things falls, as most recently demonstrated by the victory of the computer Watson over the best players in the history of *Jeopardy!*—the preeminent television quiz show. Such computer programs will perform a greater range of clerical tasks, displacing routine white-collar jobs.

The result is likely to be more unemployment in the short term and perhaps greater inequality—a recipe for social instability. Economists are right to remind us that workers who are displaced by machine intelligence need not be condemned to long-term idleness. Given the infinite variety of human desires, there is always more work to be done. But society will need to facilitate social structures that help employees face a lifetime of job changes.

Abroad, technological change will create even more disruption as the wave of acceleration engulfs societies that have not yet come to terms with the social demands of industrialization, let alone more recent technological change. Mass disorientation can become the source of both national aggression and non-state terrorism—aggression and terrorism made all the more devastating by access to weapons that are not only increasingly powerful but also deployable by ever smaller groups. Even if almost all nations in the world democratize, the one remaining rogue nation may exploit technology to cause mass destruction. Even if most terrorist organizations subside, the few that are left may gain even more power in asymmetric warfare through access to new destructive devices.

Because of such dangers, the dynamic of modern technology could as easily lead to a nightfall of civilization as to the dawn of a far better world.[2] The quality of our politics may make the difference between nightfall and dawn as we decide how to grapple with our fast-moving technological advances. Those decisions include when to regulate technologies that may prove dangerous and how to unleash from obsolescent regulation technologies that may prove beneficial. It also includes more general policy improvements to increase economic growth and social stability so that we can provide the resources and rally the popular support to address the disruptions that successive waves of technological change will cause.

The evolutionary history of mankind highlights the challenge of adjusting social governance to faster and faster change. Slowly animals evolved to learn from their environment. Through writing, Homo sapiens then became the first species to preserve learning, enabling knowledge to grow, ultimately at an exponential pace. Collective learning over time then became the source of technological improvement, a process that could move much faster than evolution.

But because human nature evolved in an era before collective learning accelerated, we are all fitted to living in a world that does not change much after our formative years. For example, the capacity to learn new languages dissipates from youth, because historically most people never leave the society into which they are born. As we come of age, we form beliefs, social networks, and allegiances without any natural expectation that they may need to be altered in light of different circumstances. During the period of our evolution, the technology that mankind created could not change, let alone destroy, the future of our children and grandchildren. Thus, there is likely a mismatch between our individual nature and the speed at which the world now changes. But while we are not individually well adapted to radical change, together we must handle the social revolutions generated by our technology. To do so requires better mechanisms for generating the kind of social knowledge that will lead to wise policy.

In the 1990s the growth of knowledge became central to explaining economic prosperity.[3] Recombining ideas in novel ways creates new ideas that allow us to do more with less and thus helps propel long-term economic growth. Just as new knowledge is a key to sustaining economic growth, it is also a key to sustaining political stability and progress. In this respect, the contemporary polity is not so different from a contemporary firm. Both must create a structure and a culture that facilitates product improvement in a rapidly changing world.

We can also compare social knowledge to the knowledge of natural science. Evidence of nature's pattern of regularities is always falling on the earth.[4] Progress in science has occurred through creating mechanisms like telescopes and microscopes to collect data and through generating a culture that seeks natural rather than supernatural explanations of the patterns generated by the data. Similarly, evidence of the social world's regularities is all around us. But we must use the best mechanisms of information technology available to gather such data and see their patterns. We also need a social culture that encourages democracy to embrace explanations that fit the data rather than explanations that stubbornly adhere to our ideological preconceptions.

The analogy to science also helps us recognize that social knowledge is always provisional. Throughout our history, scientific theories have been replaced by theories that have yielded better predictions. Yet at least since the eighteenth century, relying on contemporary science has greatly improved human life. Similarly, while new technological mechanisms can yield better assessments and predictions of social policy, such social knowledge can always be revised and improved.

The analogy also reminds us that social knowledge, like scientific knowledge, requires funds to generate it. Because technology is reducing the cost of acquiring social information, government should rationally reshape its structures and revise its laws to acquire more of it. There is widespread agreement that we should spend the resources needed to achieve such public goods as reducing pollution and avoiding damage from weapons of mass destruction. Government thus should be willing to spend resources to buy the information that helps tell us how best to furnish such goods. Social knowledge is the master public good without which other public goods cannot be well provided.

Technology progresses because humans exploit some element of the world from water and fire to magnetic fields and the quantum movement of atoms.[5] Government structure then also progresses as the new technology is deployed to create better sources of information for its social decision making, including decisions about problems created by the novel technology. Ultimately, the distinctive forms of government throughout history have been the outgrowth of the human genius for material invention, a social echo of the Promethean capture of some natural element for our collective improvement. What is new today is the pace of change and hence the pace of necessary adaptation.

A renewed agenda for a politics of learning is part of the long tradition in Western political philosophy of trying to reduce the social salience of issues like questions about religious doctrine, which have no clear answer, and focus on issues on which there can be progress, such as the creation of wealth and the extension of life. For much of the last few centuries, progress on such tractable issues has derived from the natural sciences. Today our structure of governance must create rules to elicit information from the social sciences on disputed issues of policy as well. Technological acceleration simultaneously creates a greater capacity for the polity to update policy on the basis of information and a greater need for updating to navigate the rapids of faster social transformations. The capacity of a society for learning must match its capacity for change.

## The Plan of This Book

This book comprises three parts. The first offers an analysis of the relation of social knowledge to both democracy and technological change. Chapter 1 makes the case that because of computational advances, the world is changing fast, perhaps faster than at any other time in human history. It shows how this rate of change raises acute problems for governance. It thus provides the premise for both the possibility and urgency of improving political deliberation.

Chapter 2 outlines a theory of a central function of social governance and an important function of democracy—assessing consequences of social policy—that underlies the need to create social knowledge. An inquiry into political reform today must combine what is very new—the latest developments in our advancing technology—with what is very old—the difficulties societies face in making necessary collective decisions. The rise of new information technologies has only recently begun to be explored by political theorists. Improving collective decision making by increasing social knowledge concerns matters that have vexed political theorists and statesmen since ancient Athens. But combining these two subjects offers the only prospect for political reform in our restless age of relentless change.

The second part of the book addresses four information technologies—empiricism, prediction markets, dispersed media, and artificial intelligence—each in its own chapter. New information technology can provide fuller and more objective information about policy consequences by making better use of competitive and decentralized mechanisms such as betting markets and federalism. These technologies create synergies that deliver an information stream that is more powerful than the sum of its sources. Empiricism helps us understand the consequences of past policies. Artificial intelligence can help structure the data and even suggest hypotheses that empiricism can test. Prediction markets then combine this empirical expertise with information dispersed throughout society, translating that compound into numbers that can command attention.

But these technologies can be effective only if government creates new rules for information production. The political branches must take steps to unleash prediction markets and systematically make legislation a platform for policy experiments. The judiciary must formulate a jurisprudence of social discovery, permitting states and localities to experiment in social policy, thus creating different policy regimes that can be evaluated. Government must permit dispersed media to trumpet this information about policy, particularly at election time, when citizens pay most attention.

Much of modern government is administrative government. Chapter 7 focuses on how the administrative state can integrate these new technologies to base decisions on dispersed information rather than on bureaucratic fiat.

The third and final part confronts the issue of bias. If people were so biased that they disregarded information at odds with their prior beliefs, new information technologies and new information production rules would not help in updating democratic policies. Chapter 8 surveys the many kinds of political bias. It then shows that democracy already has mechanisms to permit updating even in the face of bias. Nevertheless, bias remains a substantial, even if not insuperable, obstacle to democratic updating. Chapter 9 discusses how democracy can adopt reforms, including those based on new information technology, to combat bias more effectively.

The concluding chapter shows that our current need to enhance social knowledge is but the latest stage in the long history of exploiting technological progress to improve governance. From ancient Athens, to Britain on the cusp of the industrial age, to the founding of our own republic, successful societies have navigated the turbulences of technological change by creating better mechanisms to gather information about the social world and to translate it into wise decisions. This history underscores the fact that improving information about consequences in politics is an incremental process. It is not an argument against renovating our political information structures that reform will be imperfect or that political bias will continue to exist. The recognition that previous societies have succeeded in enriching social deliberation should give us confidence that we too can seize the opportunities newly available to begin to transform our political life.

# CHAPTER ONE

# The Ever Expanding Domain of Computation

THE COMPUTER IS NOW the fundamental machine of our age. Continuing exponential increases in computing power both propel the potentially cascading benefits and catastrophic threats that demand better governance and create the tools for better governance. The computer is the force behind most material technological advances as more and more fields from biotechnology to energy are brought within the domain of its digital power.[1] Advances in these material technologies will generate many benefits but also may create new dangers such as novel kinds of pollution and weapons.

The rapid rise of computers likely reflects technological acceleration, a process by which technological change has moved faster and faster over the course of human history. Thus, the growing power of computation will increase the pace of change, potentially generating social turbulence and instability.

Fortunately, computational advances are also driving advances in information technology, from the growth and deepening of the Internet, to the burgeoning power of empirical methods, to the increasing capability of artificial intelligence. The key to improving governance is to bring politics within the domain of such information technology. Only a politics that exploits the latest fruits of the computational revolution can manage the disruption that this revolution is bringing to the social world.

## The Exponential Rise of Computers

Computational capacity is growing at an exponential pace. Moore's law, named after Gordon Moore, one of the founders of Intel, is the observation that the number of transistors that can be fitted onto a computer chip doubles every eighteen months to two years.[2] This forty-year-old prediction has correctly foretold that every aspect of the digital world—from computational calculation power to computer memory—is growing at a similarly exponential rate.[3]

Recent studies confirm that exponential growth in computational power is still on track. In an article in early 2011 two researchers calculated that the "computing capacity of information," which they define as the "communication of information through space and time guided by an algorithm," grew by approximately 58 percent a year, very close to the eighteen-month doubling posited by Moore's law.[4] The temporal communication aspect of computation, such as broadband capacity, has grown at 28 percent per year, doubling in approximately thirty-four months. The spatial capacity for storage has grown at 23 percent per year, with a doubling time of approximately forty months.

While Moore's law is the best known perspective on the exponential growth of computing, Bell's law may better capture its social effect. Bell's law posits that computational increases of a hundredfold create new classes of computers approximately each decade. Each class of machines becomes dramatically smaller in size but with as much or greater functionality as the class it displaces. The new class then becomes a new nexus by which humans exploit computational power in everyday life. In the 1960s mainframes were a primary locus of computing. In the 1970s so-called minicomputers began their run, only to be succeeded in the 1980s by PCs. In the 1990s PCs were joined by laptops, and in the 2000s "smartphones" appeared. Networks of sensors operating together are just now beginning to collect data on such matters as traffic flows, and the ubiquitous computing of interconnected sensors will soon be upon us. By the 2020s, computers will be small enough to be routinely introduced as medical devices in the body, enabling closer interaction between humans and computational machines.[5] Already, devices like electronic tattoos are being placed on the skin for monitoring.[6]

The declining cost of computation fundamentally alters its social role. Exponential growth in computational capacity is moving computation rapidly to what technology theorists call a stage 4 technology.[7] In stage 1 a technology is very valuable and used only sparingly and controlled generally by elites. In stage 2 the technology is still expensive but cheap enough that ordinary people can use it. In stage 3 it becomes cheaper still and is integrated into every part of daily life. In stage 4 the technology becomes extremely cheap, and its use is so pervasive that it almost goes unnoticed. Computers have rapidly moved through these stages. Mainframe computers marked the first stage, personal computers the second stage, and we are now well into the third stage with mobile computing. Soon we will be in an era of ubiquitous computing when computers become so cheap they are everywhere. Computation has moved through these stages in less than a century. In comparison, paper, which began with papyrus in Egypt, took more than two millennia to become a ubiquitous presence of human life.[8]

It is difficult to overstate the power of exponential growth. As economist Robert Lucas once said, once you start thinking about exponential growth, it is hard to think about anything else.[9] The computational power in a cell phone today is a thousand times greater and a million times less expensive than all the computing power housed at MIT in 1965.[10] The computing power of computers thirty years from now is likely to prove a million times more powerful than computers of today.[11]

For years, the imminent death of Moore's law has been foretold, but the relentless exponential progress of computation has continued nonetheless. Intel—the computer chip–making company that has a substantial interest in accurately telling software makers what to expect—projects that Moore's law will continue at least until 2029.[12] It is, of course, the case that the increase in computer power based on silicon chips will not continue indefinitely, because even with recent improvements in techniques, such as 3-D chips, this kind of computing process will bump up against physical limits.[13] But in his important book *The Singularity Is Near* the technologist Ray Kurzweil shows that Moore's law is actually part of a more general exponential growth in computation that has been gaining force for over a hundred years.[14] Integrated circuits replaced transistors, which previously replaced vacuum tubes, which in their time had replaced electromechanical methods of computation. Through all of these changes in the mechanisms of computation, computational power relentlessly increased at an exponential rate.[15] This perspective suggests that other modes of development—from carbon nanotechnology, to optical computing, to quantum computing—are likely to continue exponential growth, even when silicon-based computing reaches its physical limits.[16] Although exponential growth cannot go on forever, its end is not yet in sight.

Focusing only on the exponential increase in hardware capability actually understates the acceleration of computational capacity in two ways. Computational capacity advances with progress in software as well as progress in hardware. A study considering improvements in a benchmark computer task over a fifteen-year period showed that computer speed had improved by one thousand times through improvements in hardware capacity. But computer speed had also been increased by approximately forty-three thousand times through improvements in software algorithms.[17] Like many creative human endeavors, progress in software alternates between breakthroughs and periods of consolidation where gains are less spectacular.[18] Improvements in software may also speed some tasks more than others, but in general it is a force multiplier for the gains of Moore's law.

Gains in connectivity may also increase the effective power of computation. Computers are beginning to interconnect among themselves and with human intelligence, most notably through the Internet. A concrete result of this interconnection is that more people in areas like India and China, previously remote from the core areas of innovation in the West, can now come online and contribute to the growth of science and technology. The results of the faster and greater collaboration made possible across long distances are reflected in exponential growth in the volume of scientific knowledge.[19] This knowledge creates the platform for further, faster innovation.

It might be thought that, unlike the exponential increases in computer hardware and some kinds of software, the Internet represents a onetime gain that will be exhausted when everyone goes online. But in reality the current state of the Internet represents only the beginning of deeper connectivity among machines and among humans through machine connections. First, the amount of time spent connected electronically continues to grow as connectivity is made easier through mobile devices. Second, the smaller classes of computers predicted by Bell's law will connect more and more of the physical world to the Internet, making this vast machine more sensate. Soon afterward still smaller classes of computers may also make more organic connections between the human mind and machine intelligence, rendering the idea of a global brain more than a metaphor. These interconnections are likely to further accelerate innovation. More rapid innovation requires more social interaction.[20] The denser the web of computational connections becomes, the better the innovations that are incubated there.

Science fiction writer Neal Stephenson has recently argued that our interconnectivity and informational capacity might actually decrease innovation.[21] He fears that greater opportunities for researching ideas and talking to others may discourage risk taking, because potential innovators will learn of past failures and future pitfalls more easily. But blind risk taking is unlikely to be as helpful as informed risk taking. Inventors will see what mistakes were made in the past and adapt their ideas in response, making success more likely.

The culture of improvement that has been ongoing in the West for the last thousand years also underscores the power of computational advances.[22] Once invented, almost all technologies become more efficacious as they are developed over time.[23] Given the many routes to modifying an invention, many minds can work simultaneously for its amelioration.[24] Thus, combining multiple linear improvements has often been able to generate overall exponential improvement in many kinds of mechanisms and machines.

The exponential growth of computers, however, generates broader and more powerful waves of technological change than previous inventions. Almost everything in the world can be improved by adding an independent source of computational power. That is why computational improvement has a greater social effect than improvements in previous technologies.

## Computation-Driven Technological Change

The exponential increase in computing power is driving fundamental innovation in other technological fields as they are brought within the domain of information technology. Information can be digitized and then easily manipulated, permitting the analysis and simulation of the real world without as much need for physical experiments. As a result, many fields are making enormous progress and are poised to create substantial social change. Medicine, nanotechnology, and solar energy provide excellent examples of such computation-driven breakthroughs.

### Medicine

Medicine is undergoing a computational revolution. The cost of sequencing a genome—a bounty of digital information about the particularities of individuals—is falling faster than the rate of Moore's law.[25] This progress heralds a whole new discipline of genomic medicine that will help find causes of diseases and personalize cures to make them more individually effective. Of course, it is not the case that all, or even most, diseases are wholly dictated by our genes. For instance, in recent years scientists have studied how environmental factors can affect even hereditary traits, as when cigarette smoking accelerates puberty by affecting gene expression. But such factors are now intensively studied as well. And improvements in data collection and measurement made possible by the ongoing computational revolution can help us better quantify those environmental influences. This research has the potential to create a kind of personal preventive medicine, where people would be advised to avoid the risks most dangerous to them.

Medical research will also be aided by the digitization of medical records, which will enable researchers to match genetics and personal environments with life medical histories, and by the Internet, which will allow for ease of access to such material, suitably redacted of identifying personal information.[26] Computers are becoming smaller, with the result that they will be introduced more frequently into the body as early warning systems of acute disease, as monitors of chronic conditions, and as maintenance mechanisms of well-being.[27] Thus, computational advances will lead to many different kind of innovations that will transform a central aspect of our lives—in this case, medical care.

These developments in medicine will provide benefits and yet produce social disruption—all characteristic of progress in the coming decades. As information is made more quantifiable, precise, and accessible in digital form, it will be more easily shared among researchers and become a platform for innovation, providing benefits to the public. But even beneficial improvements will create social distortions that will need to be addressed by more nimble and accurate policy changes. For instance, a rapid rise in longevity may increase the pressure on state pension schemes, forcing the reweaving of the safety nets of the modern welfare state. The ability to manipulate the genome may also create substantial risks by making it easier to create biological weapons of mass destruction. Faster transportation and globalization more generally will provide more vectors to spread engineered germs or viruses throughout the world.

## Nanotechnology

More than fifty years ago, the famous physicist Richard Feynman gave a talk titled "There Is Plenty of Room at the Bottom" in which he suggested there was much progress to be made in molding and operating matter at microscopic levels.[28] Nanotechnology—an entirely new field of research—was born soon afterward. Nanotechnology involves the use and manipulation of objects between 1 and 100 nanometers in scale, essentially at the atomic and molecular level. Nanomachines are being created to engage in such manipulation.

The fundamental dynamic of this field is also being driven by increasing power in computation. Nanoscale devices and dynamics are sufficiently small that computers are progressively better able to model and predict their behavior.[29] This modeling allows researchers to experiment with new ideas without having to endure the time and expense of creating physical models. Computers are creating virtual worlds to speed the improvements in our world.

On the agenda of this science is the creation of machines that will self-assemble. Growing computational capacity is also at the heart of self-assembly. Parts must become programmable if they are to replicate and transform themselves and must recognize their place and their geographic relation to other programmable parts.[30] If the promise of nanotechnology is realized, such self-assembly will lead to very low cost production of all kinds of industrial and household goods.

Nanotechnology will be progressively deployed in 3-D printing, another digital innovation that is already in action.[31] Here a three-dimensional printer directed by a computer program prints objects by adding materials to a virtual mold, similar to how a two-dimensional printer prints words by adding ink to a page.[32] This kind of computer driven manufacturing is already taking advantage of the connectivity of the Internet; companies specializing in 3-D printing can rapidly test crowd-sourced prototypes of a wide variety of products.[33] Nanotechnology will also have a wide variety of uses in medicine, as doctors will be able to make changes to the body at the cellular level without invasive surgery and 3-D printers will be able to reproduce body parts.[34]

But nanotechnology also brings potential costs and risks. Humans may experience some forms of nanoparticles as deadly pollution.[35] More dramatically, self-assembly may permit nanomachines, because of mistake or malevolence, to replicate without end, threatening to envelop the whole world.[36]

## Energy

Computational improvement is also creating faster improvement in the alternatives to fossil fuels. For instance, solar energy is moving along an exponential path, improving at about 7 percent a year.[37] If this trend continues, solar power is projected to become competitive with fossil fuels sometime in the relatively near future and perhaps half their cost by the 2030s. The improvements also depend on computational technology that permits the rapid design of more efficient solar cells and the production of such cells at lower cost.

One method of creating solar power deploys mirrors to concentrate sunlight so that intense heat is generated. Large mirrors are very expensive to make and maintain, but one company, eSolar, has created a system of small mirrors whose positions can be monitored and changed by computer chips in each mirror, making for both cheaper fabrication of mirrors and yet more efficient concentration of sunlight.[38] This invention provides another example of how computer chips will add value to all kinds of products in the coming years.

Such innovations also complicate collective decision making, because our decisions about how stringently to regulate today may depend on our assessments of the technology of tomorrow. If we are confident that solar energy will be available soon, government might need to intervene less to make fossil fuels more environmentally friendly. In short, the future shape of technology becomes increasingly relevant to our current decisions, because the speed of technological change is accelerating.

All of these foreseeable developments create demand for better governance. Some may create risks to the public and require regulation. Others will likely lead to deadly threats from abroad, requiring a foreign policy and defense response. Still other technologies will develop so rapidly that projecting their course will bear on current regulatory decisions. Government policy can also make a difference to the rate of beneficial innovations, and in an age of technological acceleration, even marginally faster introduction of extremely beneficial inventions can lead to substantial advances in human welfare.

## The Case for Technological Acceleration

The continuing exponential growth in computational capacity points to a more general phenomenon: technological change is moving faster and faster. In particular, the extremely rapid rise of computers to become a stage 4 technology as well as their capacity to create other technological revolutions across a wide swath of economic and social life suggest that technological transformation is happening faster than ever before.

But the case for technological acceleration is strengthened by understanding the exponential increases in computational capacity as part of a longer-term trend. Over the course of human history, epochal change has sped up: the transitions from a hunter-gatherer society, to an agricultural society, to the industrial age each took progressively less time to occur.[39] Anthropologists suggest that for one hundred thousand years members of the human species were hunter-gatherers.[40] About ten thousand years ago humans made a thousand-year transition to an agricultural society that then lasted almost nine thousand years.[41] With the advent of the industrial revolution from 1750 to 1950, the West transformed itself into a society that thrived on manufacturing.[42] Since 1950 the West has been rapidly entering the information age. Each of these completed epochs has been marked by a transition to substantially higher growth rates.[43] In addition the period between each epoch has become increasingly shorter, providing support for extrapolation to even faster transitions to a more prosperous future.

Recently, historian Ian Morris has confirmed this accelerating trend of human development, rating the level of development throughout the last fifteen thousand years through objective benchmarks such as energy capture.[44] His graph of these benchmarks shows relatively steady growth, when plotted on log-linear scale, but in the last one hundred years development has jumped to become sharply exponential.[45] Morris concludes that there may be four times more social development in the world in the next one hundred years than there has been in the last fourteen thousand.[46]

Ray Kurzweil has dubbed this phenomenon of faster technological and social acceleration "the law of accelerating returns."[47] Seeking to strengthen the case for exponential change, he has looked back to the dawn of life to show that even evolution seems to make transitions to higher and more complex organisms faster.[48] He has also suggested that the period between extraordinary advances, such as great scientific discoveries and technological inventions, has decreased.[49] Thus, even outside the great epochs of recorded human history as well as within them, the story of acceleration is much the same.

If the case for technological acceleration is strengthened by an objective consideration of the long arc of human history, it may also gain force from a more subjective look at the contemporary world. Technology changes the whole tenor of life more rapidly than ever before; at its most basic level, it changes products faster.[50] A visit to an electronics store, or even a supermarket, reveals a whole new line of products every two years, whereas someone visiting almost any store between 1910 and 1920—let alone 1810 and 1820—would not have noticed much difference in merchandise from one year to another. Even cultural generations move faster. In the space of a few years, Facebook has changed the way college students relate to one another,[51] whereas the tenor of college life would not have seemed very different to students in 1920 or 1960.

Our current subjective sense of accelerating technology is also backed by more objective evidence from the contemporary world. World growth rates are accelerating: the world economy now grows at 4–4.5 percent a year, compared to 2 percent growth in the nineteenth century and negligible annual growth before industrialization.[52] Accelerating amounts of information are being generated.[53] Information is a proxy for knowledge. Consistent with this general observation, we see exponential growth in practical technical knowledge, as evidenced by the rise in patent applications.[54] So the combination of data from our present life together with more sweeping historical and technological perspectives makes a powerful case that technological acceleration is occurring.

It is not an airtight case, however. It could be argued that the technological inventions in the 1800s, such as the railroad, or in the 1900s, such as the invention of antibiotics, marked a greater change in people's lives than we experience today.[55] The railroad, the car, and the airplane may rival any single contemporary invention with the exception of the Internet itself. But what is striking today is the pace of inventions across a huge range of enterprise that is being driven by the ongoing computational revolution. Moreover, most of these prior inventions did not have the ability to improve themselves at exponential rates. A car today is better than one made a hundred years ago, but not nearly to the degree a computer is better than its counterpart of even fifteen years ago.

Thus, the argument by entrepreneur Peter Thiel and others that the relatively small improvements in contemporary transportation undermine the case for technological acceleration may be misplaced.[56] Technological change speeds up through the discovery of new technologies like computing that are capable of more rapid acceleration. Moreover, if transportation is understood as an aspect of the more general function of making more and better connections among people, the computer revolution has been accelerating that function far beyond the progress made possible in the age of steam and rail, or of oil and cars.

Tyler Cowen's thoughtful recent book, *The Great Stagnation*, provides the most sophisticated challenge to current claims of technological acceleration in the United States.[57] Cowen argues that we inhabit a technological plateau where contemporary innovations are less salient to our lives than they were in the first three quarters of the last century. Cowen's lead argument for a technological plateau is that wages have grown much less rapidly since the 1970s than in the decades before, a claim that coincides with the frequent contention that the middle class in the United States is now doing less well.[58]

But median wage or median income growth in the United States provides a poor measure of technological acceleration for four reasons. Focusing on growth in America is too parochial, as it concerns only a single nation. Focusing on growth at the median misses the growth at the top of the income scale. Focusing on the income that actually comes to a household is too narrow, because it fails to take account of the growth in the fringe benefits, particularly from health insurance. Finally, income measures themselves do not now capture substantial increases in well being, because metrics of inflation do not reflect improvements in the quality of technologically driven products, like computers, and services, like the Internet. Each of these points is worth extended discussion.

First, median wage growth in the United States may provide a poor income measure for technological acceleration from a global perspective. Growth in world income has continued to increase. Indeed, one key story of the last decade has been an unprecedented decrease in global poverty, accelerated by strong growth in the two most populous nations of the world, India and China. Even as the deep recession began in 2008, for the first time in history the absolute number and relative share of the poorest people were declining everywhere in the world.[59] This growth is itself driven by technological changes such as citizens gaining access to cell phones and other advanced technology. It may well be that the United States is enduring slower growth because of less-than-optimal social policies. We cannot thus extrapolate from one nation or even a group of nations to measure the effects of technological acceleration around the world as a whole. A lag by the United States or other Western nations may be yet another argument for improving their information structures for fashioning policy.

Second, even within the United States, the growth in average, or to use the more precise term mean, per capita income would provide a better proxy than median household wages, because it does not focus on the distribution of the fruits of technological acceleration. Average incomes have increased far more than median incomes, because higher-income individuals have done far better in the past few decades than the average family.[60] Median family income and median household income rose only 38 and 25 percent, respectively, since 1967, but GDP per capita rose 120 percent in the same period.[61] (The graph in the appendix shows the large differences between median and average income growth and indicates that growth in average income has not been slowing.[62])

The increase in income inequality may well be a social problem, but it is more a reflection of technological acceleration than evidence against it. Because technological improvement allows relative superstars within fields to serve an ever bigger market, it has boosted the incomes of the most talented and educated at the expense of the less talented and educated.[63] Moreover, households have become smaller during the measured period, in part because of technological advances in birth control. This shrinking of family size also has a distributional effect, depressing household wages compared to the increases in per capita income.[64]

Third, the growth in median wages also does not reflect the growth in fringe benefits. Given the large increases in corporate outlays for health insurance in recent years, increases in total compensation for employees have outstripped wage gains.[65] Cowen argues that these have not been real gains in well-being: life expectancy in the United States is no better than it is in other industrialized nations that spend much less.[66] But life expectancy is depressed here by social and demographic factors that are not found as often in other developed nations.[67] For instance, the life expectancy of Asians in the United States is often higher in the United States than in their home nation.[68] Moreover, survival rates from some cancers and other diseases are also higher here, suggesting again that the citizens enjoy benefits from higher health care spending.[69] Americans may well be benefitting from health care innovations that they get before they then trickle down to the rest of the world. Without the innovation in the United States paid for by our higher costs, people here and around the world would be worse off.

Life expectancy also does not take into account the quality-of-life improvements that contemporary medicine offers, such as replacement joints that allow an active life long into old age. As former Harvard University president Larry Summers noted in a television interview, it is not at all clear whether one would choose the standard of living in 1950 with today's health care or the contemporary standard of living with health care as it existed in 1950.[70] That observation suggests that improvements in health care have made a huge difference to well-being, itself providing evidence of technological acceleration.

Even if median wages in the United States were the correct measure of income growth, they are at best a proxy for increases in well-being. A better measure would focus on increases in wealth (that is, total assets owned) that Americans have enjoyed. Adjusted for inflation, median family net worth has gone up from around $52, 000 in 1983 to over $120,000 in 2007. Average family net worth, which provides a better measure of the gains created by technological acceleration, has gone up even faster; adjusted for inflation, family net worth rose from around $137,000 in 1983 to over $556,000 in 2007.[71] Other researchers have suggested that reporting errors understate the consumption gains of individuals.[72] Americans live in bigger houses than ever before: the average home in 2000 was more than a third as large as it was in 1970.[73]

Fourth, technological acceleration may be distorting measures of economic growth by transforming what one dollar can buy. To see whether inflation is increasing, the prices of the same basket of goods must be measured every year. But economists have suggested that inflation measures are overstated in part because of failure to take into account quality changes and the introduction of new products.[74] If technological acceleration is introducing new products and improving their quality faster than ever before, the consumer price index is likely to become more and more inaccurate, leading to the understatement of economic growth.

These claims are more than theoretical. It took more than fifteen years to add cell phones to the price index's basket of goods, and that failure resulted in substantially overstating inflation in the telecommunications sector.[75] Computers are getting better all the time, but while the consumer price index tries to calculate these changes, it does so in relatively mechanical ways that fail to capture the improvements in the quality of life that such products generate.[76] I find one personal example striking. In 1973 my father gave me an HP-35 calculator costing four hundred dollars (about two thousand in today's dollars). The new smartphone I purchased this winter downloaded an essentially equivalent scientific calculator for nothing. Another example is the capacity to listen to one's chosen digital music anytime, anywhere, cheaply or even for free, without being tethered to the cumbersome stereos of yesteryear.

Access to the Internet may be the most transformative good of all. Americans spend at least thirteen hours per week online.[77] The amount of time spent may be leveling off, but this pause has perhaps occurred because search engines have become more effective, and because respondents to surveys may now have a very different idea of what it means to "go online" in an age when mobile devices such as iPads and Blackberries connect us via e-mail and social networking applications twenty-four hours a day. The unprecedented interconnectivity of the Internet is a huge boon to people in their daily lives both at work and at play, and yet it costs very little, making it difficult to compare our material situation to what it was like before the Internet. To be sure, some of the claims made for the value of the Internet—that people would rather surrender a million dollars than give it up—are overblown. But the Internet as well as iPods, mobile phones, and mobile computers are inventions whose continuing improvements have not been well captured by the consumer price index.

Finally, the apparent stagnation may well be an illusion, because computation and digitization are just beginning to spread out across disciplines like medicine, energy, and robotics. As a result, the exponential changes occurring in these technologies are not yet transforming everyday life, although they will do so soon and quickly. It is useful to remember the legendary example of exponential growth in which the sultan promised to reward his vizier by doubling each month the grains of wheat on each of the sixty-four squares of a chessboard.[78] The first doubling does not make a very visible difference. But as the doubling continues, the acceleration of growth in the amount of wheat becomes readily apparent (and dangerous to the long-term health of the vizier). In a range of technologies from biotechnology to nanotechnology, we appear to be on the fifth or sixth square of the board.

In any event, Cowen acknowledges that he expects technology to take off again.[79] His book is best interpreted as questioning the short-term constancy of technological acceleration within the United States rather than its long-term arc. Regardless of whether technological acceleration is punctuated by periods of relative stasis or presents a more continuous takeoff curve, we must prepare for future acceleration now, because we do not know when it will come.

## Social Remedies for the Problems of Technological Disruption

Technological acceleration increases the urgency of improving collective governance, because it may cause diffuse but serious social problems in addition to the particular dangers created by specific technologies. One such serious problem is the damage that technological acceleration may do to the very institutions that can help people deal with the social dislocations that such acceleration causes.

The accelerating tempo for the world today is likely to undermine important mediating institutions that have helped people cope with dislocation and social strain. For instance, for thousands of years religions have helped people through times of stress in their lives, tempering the instability caused by natural and man-made disasters. But religions depend for their strength on long-standing norms that can themselves be disrupted by technological acceleration. Recent changes in reproductive technology, such as birth control, have encouraged new norms about sexuality that are in tension with the core teachings of many religions. Accelerating technology is likely to create even more conflict with religious norms as biotechnology permits cloning and radical life extension.[80] Thus, mediating institutions like religions, which often help contain social disruption, may themselves become its source as they face internal crises of coherence in adapting old principles to a rapidly changing world.

The disruptive effects of institutions that fail to adapt is illustrated by Islam. Islam generally flourished far from the sources of industrialization. As a result, even now as technology accelerates faster with the intensification of the information age, Islam in some of its forms is left with structures and doctrines that are not well adapted to the industrial age, let alone to the information age that is being born. Globalization brings many religions into contact, mixing people of different religions on a daily basis and making it much more possible to conceive of choosing a religion other than the one into which one was born. Yet some forms of Islam take a hostile position to relations with other faiths, at times prosecuting people for apostasy. Changes in work brought about by technology have resulted in a world where women, even if generally physically weaker, can be as productive as men if they are allowed to cooperate and compete in the workplace. Yet some forms of Islam restrict educational and employment opportunities for women. In these forms women are restricted not only in places of worship but also more generally in society—restrictions best symbolized by the headscarf and even more confining garb. The doctrines of some forms of Islam are in tension with the norms needed by any modern society where men and women attend higher education together and collaborate at work.

As a result of the failure to adapt, some forms of Islam fail to help people deal with stresses of social change. They have instead become a source of instability in the modern world, as their adherents feel they are losing out to forces they cannot control. Groups with political aims are certainly exploiting Islam for violent political and social ends. But some forms of this faith can be so easily exploited precisely because they have not adjusted its doctrines to reflect the technological transformation of the world.

Islam provides a warning of the difficulties that other mediating institutions, including Christianity, may face in adapting to even faster change. Moreover, the radicalization of some forms of Islam suggests that failures of adaptation will become more dangerous as technology creates more weapons that can be effectively used by small groups as well as religiously inspired theocracies. Far from aiding social governance, some mediating institutions may contribute to the disruptions created by technological change, making better governance from the state all the more essential.

Yet the faster social tempo of the world has also contributed to making the structure of government less able to react to disruptions. One of the most troubling trends of governance in the West is the diminishing portion of the budget available for spending on new projects. The Steuerle-Roeper index of fiscal democracy measures the percentage of revenues not allocated by previously elected officials to mandatory programs like Social Security and Medicare, and the share has been decreasing over time.[81] This decrease has been due in substantial part to technological improvements. Because of better medical care and nutrition that is driven by technological innovation, people live far longer than they did at the time Social Security was conceived in the early 1930s. The result has been that a larger share of the federal budget is dedicated to old age pensions. Similarly, the increasing effectiveness of medicine in improving people's quality of life as well as longevity means that past decisions to provide Medicare and Medicaid put progressively larger portions of the federal budget on autopilot. The result is a less flexible government and one that is less able to meet new challenges with new programs.

But precisely because of technological acceleration, the federal government needs to be more flexible and innovative. Growing machine intelligence, for instance, may demand new educational programs on a large scale for faster retraining of workers. The defense budget may need to grow in response to new threats of weapons of mass destruction.

One way to respond to a less flexible federal budget is to improve our social knowledge. Better decisions on how to allocate funds will maximize benefits of the portion of the federal budget that government can currently allocate. And by bringing to bear substantial information on alternatives to some of the past allocations that are no longer cost-effective, better social knowledge can help us revisit and reverse some of these previous decisions that are putting the government into a budgetary straitjacket.

More generally, better social knowledge can help society address the social disruption that technological change brings in its wake. More efficient solutions—more effective policies to pursue the social consensus for better education and economic growth—should expand resources to address technological disruptions that may come more quickly. In addition, solutions that are seen to achieve such consensus goals will contribute to social harmony, an important consideration for an age in which that harmony may be sorely tested by disruptive social changes.

The need for better social governance certainly does not imply that government programs or actions necessarily will prove the best solutions to domestic problems created by accelerating technology. Market innovations may often provide the best self-correcting mechanism, permitting people to adapt to a world of radical change. For instance, economist Robert Shiller has suggested that because rapid societal transformation can depress wages in some sectors, workers might want to purchase wage insurance to protect against the risks of technological change.[82]

Nevertheless, even market solutions will depend on the absence of government prohibitions and may be advanced by government policies that encourage them. Political support for market mechanisms is more likely if it rests on social knowledge. Certain issues, such as pollution and weapons of mass destruction, cannot be addressed successfully by markets alone. Thus, regardless of our relative confidence in the market or in government, technological acceleration demands better governance.

The information revolution is already transforming many institutions outside of government. Every industry that touches on information—book publishing, newspapers, and education, to name just a few—is undergoing revolutionary changes as new technology permits delivery of more information more quickly and at lower cost. The same changes that are creating innovation in such private industries also can begin to renew our mechanisms of governance.

The difference, however, between information-intensive private industries and political institutions is that the latter lack the strong competitive framework for these revolutions to occur spontaneously. Hence government needs a blueprint for transforming itself into a better instrument of social learning through the integration of new technologies. Four information technologies—empiricism, prediction markets, dispersed media, and artificial intelligence—can be at the heart of such a reformation. But before turning to the description of how each technology can help create more social knowledge, we also need a theory of how social knowledge can improve political governance, particularly governance in a democracy.

# CHAPTER TWO

# Democracy, Consequences, and Social Knowledge

DEMOCRACY SERVES MANY FUNCTIONS. It helps capture the preferences of citizens, making the government responsive to what the public wants. Over time responsiveness has become a crucial source of legitimacy for government. It is not enough for a government to reflect the preferences of citizens, however; it has to be perceived as doing so. Thus, on Election Day the public display of the results of changing preferences is as important as the election results themselves. But there is a third, just as important but often neglected function of democracy: its capacity to assess and predict the consequences of social policies. It is this function of democracy—consequentialist democracy—that the tools of the information age can most easily improve.

## Consequentialist Democracy

While modern political theorists rarely emphasize the consequentialist aspect of democracy, this function has been important since its early days. In his famous Funeral Oration, Pericles, the Athenian statesman, defended democratic deliberation by pointing to its capacity for assessing the probable consequences of governmental decisions in advance: "We Athenians, in our own persons, take our decisions on policy and submit them to proper discussions: for we do not think there is incompatibility between words and deeds; the worst thing is to rush into action before the consequences have been properly debated."[1] Democracy empowers citizens to evaluate the debates about consequences by giving them a vote.

It is true that many important details of policies are left to leaders and experts to determine and to implement. But the essential roles of leaders and experts in a democracy do not lessen the need for richer information. The execution of policy by leaders and experts will also improve with social knowledge. Indeed, in many modern conceptions of democracy the functions of experts and leaders are the essence of the democratic system: ordinary people merely choose among different elites and leave them to make most of the key decisions.[2] Conceived in this way, democracy has a consequentialist aspect both because citizens evaluate elites they choose on the basis of the consequences of elites' decisions and because elites take consequences into account in making their decisions. It also creates a complex system of information feedback in which experts generate information that helps citizens evaluate which policies are most likely to realize their values and objectives.

This function of democracy takes values and objectives as largely given. For example, consequentalist democracy in the United States takes it as a given that the government should act to improve the educational outcomes of children. The key question of how to do so depends on the consequences of various plans. Consequentialist democracy also has an important factual premise of its own. It assumes that enough people in a society share sufficient objectives that many questions are not simply a debate about which values should be chosen but about the means to realize a set of consensus objectives.

The United States in fact has sufficient consensus to make consequentialist democracy productive in many areas of social policy. First, there is a consensus that collective decisions are necessary in some areas. It is widely recognized that the United States must make collective decisions about certain matters such as national defense and pollution. The goods at issue—security and a clean environment—are public goods that cannot be sufficiently supplied by the market without public intervention. For example, because everyone enjoys the benefits of national defense and no one can be excluded from that enjoyment, market transactions will not supply enough national defense. Clean air and other environmental amenities that are enjoyed in common are similar. Without government intervention, companies may unduly pollute the air.

Of course, this consensus on the general need for collective decision making does not mean that citizens in general can give an economic explanation of what constitutes a public good. But they agree generally that government's role should be focused on goods that the market and family cannot sufficiently provide, although there is room for disagreement about what those goods are. In addition to this very general agreement about the purpose of government, many specific public goods themselves enjoy a substantial consensus. Sustaining good education for children and promoting economic growth are instructive examples to explore in more detail.

While education, both primary and secondary, has benefits to individuals, it is widely thought to benefit society as well.[3] Education helps make people more productive. It makes them less likely to be charges on the public treasury. It enables them to participate more fully and intelligently in democratic decision making and thus is itself part of the feedback system of consequentialist democracy. So, while some benefits of education redound to the individual, some also redound to society. As a result, if government did not aid in its provision, education would not likely be produced to the degree that its total benefits (which include both individual and social benefits) warrant.

Of course, exactly how government aid is to increase educational outputs is a matter of intense contemporary debate. Will vouchers or charter schools improve attainments? Even within government-run schools, will merit pay or smaller-size classes or both raise test scores? But these important questions are ones of means, not ends. Education is what political scientists call a valence issue, in which most people largely agree on the objective even if they disagree on the policy instruments to reach it. Consequentialist democracy focuses on the best means to meet common goals.

Economic growth has some aspects of a valence issue.[4] President George W. Bush proclaimed that tax cuts for individuals offered a way out of recession and would increase economic growth.[5] President Barack Obama proclaimed that his stimulus plan of higher government spending offered a way out of recession and would increase economic growth.[6] These were very different plans from presidents of different parties, but the proffered objectives were broadly similar. As to issues of growth, what is debated is which political program will broadly deliver these economic goods.

Consequences matter even on issues within that debate where there is disagreement, as, for instance, on the issue of balance between growth and equality. Most people favoring more income equality will nevertheless want as much growth as is consistent with an equality constraint; most people favoring growth will nevertheless want to minimize harm to equality. Indeed, characteristically politicians argue that their policies will deliver both. President Obama argues that the "investments" that the government will buy with additional taxes will pay off in jobs and prosperity. President Bush suggested that income tax cuts will allow small businesses to expand, creating jobs and helping the less advantaged in the long run through greater economic growth.

Thus more information about the consequences of various economic policies will make it easier to find the best way to increase economic growth without substantially reducing social equality, or the best way to reduce economic inequality without substantially decreasing economic growth. Consequentialist democracy can therefore make progress on issues that are part valence and part value issues. Indeed, given that the amount of trade-off, if any, between equality and growth over a substantial period of time is a question of fact, this trade-off itself can potentially be evaluated by consequentialist democracy.

One reason that citizens' objectives often converge is the threat of competition with other nations and foreign organizations.[7] A modern nation-state needs strong economic growth, if only to be able to compete in the military sphere. As a result, institutions that helped generate economic growth, such as the General Agreement on Tariff and Trade (GATT), met with widespread support during the Cold War because they empowered the West vis-à-vis the Soviet Union.[8]

Today the existential threat has retreated to some degree, but the perception of geopolitical competition among nations sustains the consensus on the need to improve education.[9] For instance, the rise of China prompts widespread concern about how to make sure our children have the education to keep our nation's economy competitive. In part, Americans want to make sure that the United States has the resources to continue to occupy a position of strength in its dealings with this rising power.[10]

The set of issues more amenable to political resolution through investigation of consequences is not fixed, but expands with the increase of social knowledge. When an individual finds it hard to evaluate the consequences of a social decision, he or she is likely to evaluate it by reference to some diffuse attitude already held, such as fear of government or envy of the rich. Insofar as more information can be gained about the actual effects of a policy, some citizens are more likely to be influenced by those consequences in their decisions. As a result, the number of issues amenable to factual resolution in the political realm is related—or, as social scientists would say, endogenous—to the growth of social knowledge.

Additional information is unlikely to help resolve disagreements on some political issues such as abortion.[11] But democratic factual updating can still be helpful so long as there are a variety of topics where greater convergence on the facts generates greater political consensus. For reasons I will discuss in chapter 8, democracy makes a lot of mistakes, adopting polices that do not achieve the goals for which they are designed. But new information technologies can help us reduce the number of those mistakes by providing more information that will energize the forces in society focused on promoting commonly shared public goals.

## The Power of Reducing Information Costs

Understanding the consequences of social decisions requires the use of social knowledge—that is, true information about the world. Democracies also work more effectively when basic social knowledge is more widely shared, because at election time citizens must rely on information to assess whether the proposed policies of their leaders are broadly sound. Modern information technology can facilitate acquiring social knowledge by reducing information costs. Reducing information costs has four large advantages for social decision making. The first is the most obvious: reducing information costs can create more knowledge about public policy and reduce the cost of accessing knowledge. The second advantage is more dynamic: reducing information costs better enables citizens to organize around encompassing interests that many of them have in common, such as good education and economic growth. Third, reducing information costs makes it more likely that citizens will vote on the basis of these encompassing interests rather than their own private interests, because it makes them more confident about which policy instruments will achieve such interests. Finally, reducing information costs helps to better mix the information from experts with the dispersed information of the general public, thereby making social knowledge more accurate and helping the democratic agenda to focus on the most salient issues. Lowering information costs thus not only enlarges social knowledge but also creates a knowledge dynamic that provides citizens with more incentives and more capacity to use that knowledge to pursue interests that they have in common. These four effects are important enough to describe in detail.

### Creating Common Knowledge

First, reducing the costs of creating and accessing information creates more knowledge about public policy and makes more of that knowledge common. The growth of common social knowledge itself can potentially help policy makers make decisions that reach consensus policy goals and avoid pursuing programs that may seem on the surface beneficial, but are in actuality counterproductive.

The information in consequentialist democracy is information about facts. The relevant factual propositions include both factual foundations for policy, like the extent of global warming, and possible policy results, like the effects of a carbon tax on global warming. A focus on revealing and refining factual knowledge differentiates consequentialist democracy from other possible functions of democracy, such as its capacity for eliciting preferences,[12] or its capacity for changing preferences through deliberation about values.[13]

The preference-eliciting function works best in dialectic with the consequentialist function. We need to elicit preferences in order to formulate objectives for policies. Consequentialist democracy then attempts to evaluate the results of these policies. Such evaluations may show that policies that are supposed to achieve the preferred goals can be counterproductive or have unintended effects. This kind of sifting may often lead to new policies that attempt better to achieve preferences. Sometimes repeated policy failures may lead to the reevaluation of preferences, because they may suggest that the preferences cannot be achieved or can be achieved only at too high a cost. The costly failures of nationalization of industries in Western Europe led many even on the left to abandon such government control as a desirable goal.

It is less clear that consequentialist democracy benefits from deliberation to transform values. Some see such deliberation as too abstract an enterprise to command the attention of citizens. Others believe that the political fragmentation resulting from a debate about very divergent values undermines political parties and their capacity to articulate issues of broad concern.[14]

Whether or not democracy is a useful forum for value clarification, consequentialist democracy is not subject to the same criticisms. Because policy issues and their evaluation involve concrete matters, such as what tax regime will help economic growth or what structure of schooling will provide students with a better grasp of math, consequentialist democracy is a less abstract enterprise than most conceptions of deliberative democracy. Because consequentialist democracy focuses on evaluating policies that are seriously proposed or enacted, it does not fragment the polity, but instead focuses citizens on an evaluation of programs with substantial support. In short, consequentialist democracy may sometimes change the preferences of citizens, but it does so through focused deliberation on programs rather than abstract deliberation about values.

One reason for caution about the beneficial effects of greater social knowledge is summed up by Alexander Pope's famous lines: "A little knowledge is a dangerous thing / Drink deep or taste not the Pierian spring." Partial information might be thought sometimes to mislead the public. For instance, transparency about the salaries of public sector employees may simply prompt them to take more of their compensation in the less transparent form of pensions or tenure protections. But the easier it is to generate fuller information and to provide a structure to make it comprehensible, the less likely it is that any additional amount of information provided will prove counterproductive. Today modern information technology provides the means to calculate and then to publicize the exact value of pension benefits and to compare tenure protections in various posts. Thus, it is unlikely that public sector unions and other special interest groups will be as easily able to evade the constraints of transparency as information costs less to access. A harder question about the value of increasing information is whether more knowledge will translate into better decisions about such matters as economic growth and better schooling. To be effective, policy makers must use new and accurate information as a basis for decisions to achieve encompassing interests. They are much more likely to do so, however, if the creation of more knowledge also rallies citizens to encompassing interests.[15]

## Creating Incentives to Act on Encompassing Interests

Lowering information costs can provide incentives for political entities to enact better policies by permitting citizens to more effectively pursue encompassing interests. Most people do not act only in their narrow self-interest in politics, but they also have moral sentiments and an inclination to sociality.[16] Thus, even without good information, citizens have some tendency to promote encompassing interests.

But citizens also face inherent difficulties in organizing for such encompassing but diffuse interests as faster economic growth or better education, because they face collective action problems. Since everyone will benefit a relatively little bit from the common goal, no one has sufficient incentives to pay the costs of pursuing it. Obstacles to organizing around encompassing interests include discovering what those interests are, determining what policies will promote them, and finding like-minded citizens with whom to pursue the policies—all problems that can be ameliorated by lower information costs.

In contrast, groups with concentrated interests have less difficulty organizing around those interests even in a world with higher information costs, because the return they can get from organizing is more substantial. As a result, citizens more easily come together to engage in seeking particular benefits from the government (called rent-seeking in the economic literature), as when sugar growers today find a common interest in lobbying for subsidies.[17] In a world of high information costs, special interests are likely to dominate collective decision making, seeking results that are beneficial to themselves rather than to society more generally.

For at least the last half millennium, information technologies have played an important role in changing the balance of power between special and more encompassing interests, because greater capacity to access and distribute information is generally more advantageous to encompassing interests. The "technologies of freedom," first in the form of the printing press and subsequently in the form of weekly or daily newspapers, helped the rising middle class in Western Europe and in America to democratize their governments.[18] In the first part of this period, the special interests were the rulers themselves and their supporters, who ran the state largely for their own enrichment. But as the costs of information fell, the middle class was able to find encompassing interests around which to organize, like the slow expansion of the franchise, in order to counteract the special interests that ran the regime.

To be sure, a catalyst was often necessary to curtail the power of rulers and their exactions. Sometimes it was an event around which people could coordinate—for example, the British Parliament's Stamp Act of 1765. Alternatively, a charismatic leader like Oliver Cromwell brought people together, or a religious movement like Protestantism appealed to powerful transcendent values to overcome collective action problems. Nevertheless, the amount of such collective action at moments of crisis depends on the costs of discovering and organizing around encompassing interests—costs that depend on the deployment of information technology of the day. The slow reduction of these costs over time makes it easier to harness more diffuse moral sentiments, because people can better appreciate the concrete steps to translate these sentiments in action.

It is thus hardly a surprise that rulers often attempted to prevent effective use of the information technologies of the day—the threat that our First Amendment is designed to prevent. Even in our age, repressive regimes like those in Iran or Syria shut down the Internet and Twitter at times when these regimes are threatened. However, even after societies are democratized, the struggle between encompassing interests and special interests continues in new, if less dramatic and violent, forms.

In a mature democracy, special interests are not monarchs or aristocrats, nor are they even principally the elected leaders themselves. Instead they are groups like unions and trade associations that are able to overcome collective action problems and wield substantial influence over government policy because of their concentrated power. Special interests helped create a complex tax code full of exemptions and preferences that serves concentrated groups at the expense of economic growth, a more encompassing interest. By blocking reforms, teachers unions can thwart increasing output in public education, another encompassing interest.[19] Other self-identifying interest groups form around government programs that benefit them; for example, the elderly organize around Social Security and Medicare, impeding beneficial reforms. It is an axiom of democratic politics that legislatures tend to provide concentrated benefits to such groups and impose costs on the more diffuse citizenry.

Social complexity has also led to an increase in the number of government programs. The result is that special interests have more opportunities to gain benefits, and the diffuse citizenry has a more complex array of policies to monitor.[20] Special interests routinely exploit this development by raising information costs: they increase the opaqueness and complexity of government programs. They portray themselves as guardians of the public interest even as they exploit the public treasury.[21] For instance, teachers unions can argue that higher pay raises across the board will improve education. Or oil companies can argue that giving them tax breaks will reduce the price of gasoline. Such claims may be false, but their specialized nature makes it difficult for the public to investigate and assess them.[22] Special interests regularly oppose transparency, as when they fight against the disclosure of campaign contributions. Special interest opposition to transparency itself provides perhaps the most compelling evidence that greater information can help encompassing interests.[23]

Just as the technologies of freedom like the printing press helped the middle class to discover and act on a singular and focused encompassing interest in more democratic and accountable government, so our new information technologies can help citizens to do a better job of mapping our far more complex policy landscape by reducing information costs still further. Citizens can then more easily discover policy instruments to pursue encompassing interests, and better understanding can give more incentives to organize around common goals. To be clear, lowering information costs does not dissolve special interest groups; they will continue to lobby on behalf of more narrow causes. Nevertheless, reducing information costs can slowly change the balance of power between special interests and groups focused on more encompassing objectives in society.

Technological change is also likely to create exogenous shocks that may help the new information to be more effective in changing policy toward that favored by encompassing interests. Technological changes make it harder for many kinds of special interests to retain their power. For instance, the Internet offers ubiquitous online learning, undermining the local leverage of teachers unions.[24] Moreover, rapid change makes it more difficult for politicians to enter into implicit long-term deals with interest groups such as corporate trade associations at the expense of the public. The exactions gained today by such a group from an inefficient regulation can be gone tomorrow, because technological obsolescence makes the regulation irrelevant. Thus, in an era of technological acceleration, citizens may face less resistance in their pursuit of encompassing interests.

American history has previously demonstrated that reducing information costs can change the structure of politics. As Bruce Bimber has shown, different information technologies have previously changed the capacity of groups to pursue common goals.[25] In the early and mid-nineteenth century, for instance, the rise of newspapers and the creation of the post office lowered information costs to allow parties to organize.[26] Thus, the new information mechanisms of our age of technological acceleration can be seen as continuing a century-long trend. The government should consider subsidizing modern information technologies to create better public policy information today just as it previously subsidized post offices. If the government is justified in purchasing public goods like defense and education, it is justified in subsidizing the technologies of social knowledge that help tell us how best to produce such goods.

## *Improving Incentives to Vote on Encompassing Interests*

Reducing the costs of information also provides better incentives to vote on the basis of encompassing interests rather than merely parochial ones. Political scientists have shown that citizens vote in part on the basis of encompassing interests; for instance, they are concerned with economic growth generally as well as their own pocketbooks.[27] Such a voting pattern even comports with a model of human nature in which people directly pursue only their own interests. Because it is very unlikely that a citizen's vote will make the ultimate difference in an election, he or she can afford to use voting to reinforce his or her self-image.[28] An important part of that image likely lies in regarding oneself as a public-minded person.

Reducing information costs tends to reinforce that more public-minded motivation, because given dual motivations, citizens may still focus substantially on parochial interests insofar as it remains unclear how common interests will actually be achieved. For instance, individuals have generally had more reason to believe that a subsidy would benefit them than that a policy would lead to economic growth. The concrete benefits of a subsidy were obvious, whereas knowledge about growth-enhancing policies was sparse and uncertain. Insofar as the policy instruments to achieve a consensus objective become more widely and more certainly known, citizens become more confident that their views will actually redound to the public interest. This confidence encourages them to emphasize their inclination to vote on the basis of encompassing interests. In short, social knowledge gives wing to the better angels of our nature.

## *Mixing Expert and Nonexpert Opinion*

Lowering information costs also makes it easier to mix expert and nonexpert information to create better and more useful social knowledge. Creating such a mixture has been a goal of democracy since ancient Athens.[29] A mix of expert and nonexpert information on any subject is likely to be better than the best expert information or the unfiltered popular sentiment. Experts bring a deep base of knowledge to any policy issue. With this ballast they are less likely to be swayed by the latest information or passing fad.[30] But experts can be insular and subject to the biases of their class. It is possible, for example, that bias may infect some social sciences where most experts today are generally ideologically liberal. It is difficult, of course, to ascertain the extent of any ideological bias. Perhaps social scientists' knowledge has moved them to ideological liberalism because, at least in their area, liberal policies have better outcomes for human welfare than conservative ones. Nevertheless, it is impossible to be confident that their positions are not affected by bias, particularly since experts have some incentives to conform to the majority opinion in their social and professional circles.

Nonexperts also have some advantages in assessing social policy. They are more numerous, and thus they enjoy the accuracy that comes from the wisdom of crowds.[31] They are more dispersed and see the policy from different angles. They are less insular and, collectively, may be less biased at times. In short, their participation in assessing consequences helps assure that these decisions will reflect more dispersed information.

The need to combine expert and nonexpert information can be illustrated by predicting the results of an election. Obviously, polling is one way to gather important information about the likely results. But this information provides only a baseline for prediction. The outcome of the election will be influenced by events between the polling time and the date of the election. Information about such events is clearly distributed unequally. One reason is expertise. Some experts have studied elections and know that certain factors are likely to influence how people vote yet are not reflected in the polls. For instance, experts know that challengers with low name recognition are likely to gain support as they become better known.

Nonexperts may also have particular information that is helpful to predict the electoral outcome. Most dramatically, some may know of unfavorable personal information about a candidate, such as extramarital affairs or unethical business dealings, that is likely to come out in the course of a campaign. Others may have knowledge of local economic conditions that are likely to have an effect by Election Day. Thus, the cumulative knowledge of nonexperts may prove important.

The distribution of information about likely policy consequences among experts and nonexperts is not substantially different from the distribution in predicting an electoral result. Experts in the area often have better information about the probable results, so economists are likely better able to evaluate the consequences of a free trade deal than nonexperts. But some nonexperts have useful information as well. They might know something about local economic conditions that could affect the consequences of a policy, or they might possess information about the particular people who are charged with carrying out the policy and who thereby influence its results.

The new information technology of prediction markets is particularly helpful for combining expert and nonexpert information in a market on a particular subject. As we will see, prediction markets allow experts and the general public to bet effectively against one other in predicting policy outcomes.

A combination of expert and nonexpert knowledge not only provides advantages on particular subjects, but it also helps assure the best use of the inherently limited capacity of democracy by prioritizing the most important issues. The problem of democratic capacity is exacerbated by the problem of rational ignorance in politics. Because a citizen's vote is very unlikely to decide an election, it is not rational to spend much time to become politically informed, particularly when other uses of one's time are more lucrative or entertaining.[32] Some people follow politics for the fun or spectacle of it as others follow sports, but it is clear that these numbers are not very great. Politics for most is simply not as satisfying as rooting for a local team or watching a blockbuster movie. In contrast, experts pay close attention to what is important in their own field, but they tend to assign undue importance to issues from which they earn their living.

Thus, society needs nonexpert institutions to gather the results of expert information in various fields and bring the most important issues to the fore so that citizens can assess competing policies. Technological acceleration makes the need for such prioritization more acute, because the fate of civilization may turn on discovering the most important risks and deploying substantial resources to avoid them.

Combining expert and nonexpert information helps democracy focus on the most important issues and make decisions based on the best information available. Refracting expert information through nonexperts helps prioritize among subject matters for the triage of democratic decision making. As we will see, blogs and other dispersed media funnel expert opinion to the wider world and help to provide a more accurate mechanism for prioritizing the issues of the day.

Creating more social knowledge thus has a variety of dynamic effects, helping citizens organize and vote for more encompassing interests and creating better guides for the policy instruments to achieve these interests by generating a better compound of expert and dispersed information. It also helps citizens indirectly by making experts and leaders better informed. More social knowledge about consequences of social policy reorients democracy to focus on the social policies that can best achieve its most important consensus goals.

## Information from Markets and Information from Democracy

Democracy is not the only social system that helps evaluate consequences by pooling information. The market's price system is a system of information, a discovery machine for improving our decisions. Prices tell us the cost in resources of satisfying particular human desires; in fact, the equilibrium reached by markets provides a mechanism for satisfying those desires most efficiently. Markets generally provide much better information about the costs of individual decisions than democracy does about the cost of collective decisions. One reason is that individuals are likely to focus more on their purchasing decisions than on their electoral decisions. Their individual decisions determine what they buy. An individual is not much more likely to affect the result of an election than he or she is to be hit by lightning on the way to polls.

But the greater efficiency of the market as opposed to the government in allocating resources in most instances does not avoid the necessity of collective decision making. Given the need for public goods and the failure of markets in some circumstances, decisions about the boundaries between collective decision making and individual market decision making are themselves necessarily made collectively. These boundaries are often contested not least because the beneficence of state intervention depends on factual claims about the consequences of intervening and the alternative consequences of not intervening. For instance, the decision to have the government attempt to regulate fossil fuels depends on an evaluation of the dangers of global warming and the efficacy of regulation in tempering those dangers. Collective decision making is also necessary to cut back on excessive government. When regulations are ineffective or outdated, they need to be scrapped, and this decision too is taken collectively.

Limiting the scope of government may help maximize the scope for market decisions, but limited government is not government's negation or anarchy. However confined is government's scope, collective decisions on constitutions or other fundamental structures govern the nature of its limits. Information is relevant to determining the appropriate limits and constraints on government. Within those limits, government requires collective decision making. Thus, whatever the appropriate limitations on government, it remains useful to consider how a more informed politics might make decisions within the government's ideal scope the best they can be.

Even if the success of the market cannot replace the need for government, one important advantage of accelerating technology is to bring to political decision making some of the advantages of the market and its mechanisms of gathering dispersed information. Thus, federalism can create a market for a social governance whose results can be better evaluated than ever before because of the more sophisticated empirical methods that exponential increases in computational power support. Like contemporary corporations, the government can also conduct randomized experiments to evaluate the consequences of its initiatives. Markets on a scale made possible by the Internet permit citizens to bet on policy results, gathering dispersed information that is latent in the society. Society can thus better apply to public decision making the competitive and decentralized sources of information rightly celebrated in private markets.

## Nondemocratic Mechanisms as Aids to Democracy

Most of the mechanisms that help democracy assess consequences better in this age of technological acceleration are not themselves democratic in the sense of weighting the opinions of all citizens equally. Prediction markets, for instance, weight the views of those who are willing to bet the most on predicting outcomes, because the willingness to bet signals superior access to relevant information.

But relying on betting markets and other mechanisms of assessment that have unequal participation can still advance consequential democracy without undermining democracy's other functions. Citizens still have an equal ballot and can decide for themselves what difference the information provided by such mechanisms should make to their voting decision. Even now citizens vote their preferences in a system where experts have more influence than most people in creating the social knowledge of likely policy consequences. The consequentialist function of democracy does better when citizens possess more accurate information about policy consequences and can then choose between them on the basis of preferences. As former senator Patrick Moynihan famously said, "Everyone has a right to his own opinion but not his own facts."

More radically, economist Robin Hanson has suggested that we might create a structure in which people separately vote on the basis of preferences and bet on predictions about the consequences of policy. Such a clean separation could better reflect two desirable objectives of government: distilling preferences and predicting policy consequences.[33] But the current democratic process makes it impossible to cleanly divide preferences about policy views from assessment about their consequences. Citizens and political representatives often do not neatly distinguish preferences and consequences in their own minds, at least on issues where both factors influence their views. Moreover, in a representative democracy, citizens also vote for representatives who act both on values and their prediction of policy consequences. Finally, because legitimacy in the modern world comes from the perception that the people rule, citizens must make the decisions about which leaders to choose even as they use new structures to better predict the consequences of what those leaders will do. For the foreseeable future, new information technologies will function in democracy as adjuncts in helping citizens to determine the relevant facts rather than as mechanisms binding government to use particular means to reach consensus goals.

Given the power of technology to transform our social landscape, it is possible that a future political system will be able to disentangle registering preferences and evaluating consequences. In such a world, politics would follow a two-step process. Citizens would first vote on their preferences. But the government would then use other mechanisms, like prediction markets, to choose policies that would best achieve those preferences.

Nondemocratic regimes may be able to follow prediction markets and other mechanisms of consequential evaluation directly without engaging in the first step of voting to register preferences. If those systems become more successful than democracies at predicting the consequences of social policy, they may spur democracies to create a more formal separation between the functions of eliciting preferences and evaluating consequences. But democracies will undertake such truly radical change only if they gain experience in using mechanisms for predicting consequences as aids to traditional democratic processes rather than as replacements. It is to these mechanisms that we now turn.

# CHAPTER THREE

# Experimenting with Democracy

TECHNOLOGICAL ACCELERATION increases the capacity to measure the consequences of government action. The resulting social knowledge of which policies work can improve government performance and decrease social disagreement. Government itself can be a catalyst for increased social knowledge if government transforms itself into a better instrument for social learning by adopting rules that will better examine which government programs are successful. These include rules encouraging policy experimentation through decentralization and randomization, providing incentives for improved research practices, and making government data more transparent and accessible. The information age permits us to foster a more experimental politics.

## The Nature of Social Science Empiricism

Empirical social science attempts to discover the causes of social behavior. One cause of social behavior is social policies. Insofar as empirical investigators show how various policies affect social behavior, they create social knowledge that can improve policies on the assumption that there is a consensus on the behavior to be either encouraged or discouraged. Thus, like natural science, empirical social science seeks the causes of things. But the nature of social phenomena makes this task difficult. Experiments can be designed to isolate the causes of natural phenomena, but social science generally faces the difficulty of trying to infer causes from a welter of real-world data.[1]

For instance, suppose we want to know the effect of merit pay awarded to teachers on the basis of improvement in class test scores on a particular kind of human behavior to be encouraged: student learning. There are a variety of possible theories about merit pay. On one hand, the theory in support is that it will provide incentives for teachers to teach better, improving long-term educational outcomes for their students. On the other hand, perhaps it will encourage teachers to teach to the test, with no gains and perhaps losses to the long-term educational outcomes.

Such educational outcomes have many causes. How is one to isolate the effect of merit pay? The most obvious way is to compare graduation rates, college admissions, and other relevant effects in jurisdictions that use merit pay to those that do not use merit pay. But this inquiry does not dispose of the problem. The two types of jurisdictions may vary in other features, and these may turn out to be responsible, either directly or in conjunction with the presence of merit pay, for the difference in outcomes.

To be more specific, assume that Massachusetts had statewide merit pay but Idaho did not. Even if Idaho over time showed substantially lower educational outcomes than Massachusetts, it does not follow that merit pay is the reason for that difference. There may be another difference between Massachusetts and Idaho that is the "but-for" cause of the lower attainments—anything from differences in demography to differences in culture. Moreover, because of the legislative decisions made in each state, the observer can see the effects of merit pay only in Massachusetts, not in Idaho, and the effect of the absence of merit pay only in Idaho, not in Massachusetts. As a result, social science empiricism routinely faces difficulties in establishing causes of social phenomena, particularly the difficulty of excluding the possibility that correlations do not reflect causation.

But social scientists have found clever ways to control for this possibility. One is to multiply the number of different jurisdictions studied. By considering more jurisdictions, they are better able to discount the influence of factors unrelated to merit pay, because some jurisdictions are more likely to have similar features aside from their positions on merit pay. Second, researchers can introduce other variables to account for such possible causes, like demography or culture, and then assess the effects of these variables.[2] By so doing they reduce the likelihood that unobserved factors explain correlations.[3] To be sure, making analysis more complex may allow researchers to introduce ideologically driven distortions, but the growing culture of empiricism is likely to restrain ideological manipulation.

In some cases, social science empiricists are able to more closely approximate scientific experiments through so-called natural experiments. A natural experiment is an experiment in which a scientist finds a real-world event that changes the levels of a variable he or she wants to measure. A natural experiment depends on an event that is random with respect to the factors that may prevent correlation from reflecting causation.

Jonathan Klick and Alex Tabarrok used the trigger of Homeland Security alerts as a natural experiment to demonstrate the effectiveness of police.[4] Homeland Security alerts increase the number of police officers in a given area but are otherwise a random event with respect to crime rates.[5] Thus, the effect of varying the number of officers in an area can be directly assessed. On the basis of this information, their analysis suggests that a 50 percent increase in the number of police officers reduces crime by approximately 15 percent.[6]

Other economists have discovered randomness within events that on their surface appear to lack randomness. Comparing the performance of companies with unions to that of companies without unions is difficult, because the companies that decide to unionize may have characteristics that are different from those that do not unionize. Thus, there may be a variable other than unionization that actually explains the difference in performance. To address this problem, two economists considered companies whose employees either very narrowly voted for unionization or very narrowly voted against it.[7] For these companies, unionization was essentially random with respect to other characteristics, because events unrelated to the characteristics of the company, such as the weather or the pattern of illnesses, were decisive in electoral outcomes. By focusing on companies with such close elections, it became easier to investigate the actual effects of unionization on companies' prospects.

The government can also consciously create randomized experiments, so-called field experiments. It can facilitate social learning by assigning different individuals or groups of individuals to different programs at random. Social scientists can then measure the distinct outcomes. The idea here parallels medical trials, in which patients are randomly given a new drug or an old drug or a placebo and their health outcomes are then measured to gauge the effect of the new drug.

While natural and field experiments add to our social knowledge, they do not necessarily end the need for further inquiry. One possible limitation on such experiments is whether the results can be generalized.[8] Klick and Tabarrok's study of police effectiveness focused on the Washington, D.C., area. One might question whether their results carry over to nonurban areas and even other urban areas. Nevertheless, the experiment almost certainly tells us something valuable about Washington, and unless we can think of relevant distinguishing characteristics, it will provide useful presumptions about other urban areas. If other factors seem to distinguish other areas, we can then focus on testing the effects of these other factors. Natural or field experiments in other geographic areas can gauge whether similar results are achieved. As more such experiments are made, the cumulative effect will be to expand our knowledge of policy results.

Another limitation is that these experiments in themselves do not show the mechanism by which the policy causes the effect, and thus they do not necessarily isolate the crucial ingredient of the policy. In other words, they may show that a policy works without showing why it works. Nevertheless, at times the mechanism is pretty evident. Having more police on the streets deters criminals, because they face a higher chance of being caught. Even in cases where the mechanism is less clear, replication of all the policy ingredients may obtain similar results without certainty about the mechanism. Theories of human behavior validated by other evidence may help us understand the mechanism. Natural and field experiments do not resolve all of these questions, but they are valuable pieces of the mosaic that makes for better social knowledge.

Some people argue that empiricism does not have the potential to help settle policy disputes, because the causes that empiricists infer are not really facts, since they are not directly observed. But much, if not most, knowledge that affects how we live our lives is not directly observed. Most of us do not see that a molecule is made up of atoms, but we accept such a proposition either by inferring it from other facts or from reasonably crediting scientific authority. Knowledge about social behavior may be more difficult to discover, but it remains knowledge even if it depends on inferences and authority.

## The Rise of Empirical Analysis

The accelerating power of computers, discussed in chapter 1, addresses what has always been the Achilles' heel of social science empiricism. Pythagoras famously said, "The world is built on the power of numbers." That is the slogan of empiricists as well, but in order to carry it out successfully, empiricists need enormous amounts of data and large calculating capacity to tease apart causation from mere correlation. First, the social world must be broken down into numbers that can be calculated, and to deal with matters of any social complexity, a lot of numbers are required. To draw any conclusions, these numbers then must be sliced and diced to test hypotheses about particular social claims, such as the assertion that a certain kind of charter school improves test scores or that the deployment of more police decreases crime.

But today the increase in computational capacity makes possible more precise measurements of worldly events.[9] More data is being put online at a rapid rate through social media and other mechanisms. The coming ubiquity of networked sensors will also collect far more data about the physical and social world, making more data available on the World Wide Web.[10] Individuals are themselves creating data about all aspects of their lives and organizing it in charts and graphs.[11] Soon researchers will use electronic agents to collect and organize online data.[12]

One particular advantage of this growing capacity for gathering data is the increased ability to gauge outcomes with both greater comprehensiveness and nuance. For instance, objective measures of quality of life can supplement life expectancy in assessing health care initiatives. Job success can supplement test scores in assessing educational reforms. Moreover, data will permit researchers to find proxies to test for the effects of policy on other measures of social health. They can measure, say, whether improved methods of schooling can lead to involvement in the community. More comprehensive data can help measure the effects of policy not only on economic matters but also on more subtle matters of culture. More data also create more opportunities to find triggers for natural experiments.

Greater computer calculating power permits the construction of more complex empirical methods. When natural experiments are not available, investigators may use such equations to exclude confounding factors and reveal the causes of social phenomena.[13] Multiple-level modeling techniques require more computer power. Yet these techniques will be better at measuring social phenomena that depend on the complex interaction of actors at different levels of social hierarchy, such as when pupils are nested in classes that are nested in schools that are nested in school districts.[14] Greater computational power also permits greater use of methods like repeated sampling that give researchers more confidence in their results.[15] Such methods allow social scientists to correct for smaller sample size by running simulations and effectively obtaining better analysis based on less data than previously.

Additional well-executed empiricism can have a cumulative effect. More studies assessing the effects of social policy allow meta-analysis that combines the result of many studies. Such meta-analyses can be more persuasive than any single study, because they consider more data as well as different approaches to data. A rope is usually stronger than the strands from which it is woven.

Perhaps most important, the declining cost of empiricism also changes the culture of social science analysis by changing the relative costs of empiricism and theory. A hundred years ago, armchair speculation was very cheap compared to empiricism, because the technology available for empiricism was not sufficient to deliver any useful results. Even forty years ago it might have taken a social scientist several months to complete a regression analysis. Now participants at faculty workshops use their laptops to run regressions on the speakers' data in a matter of minutes or challenge conclusions by pulling other data sets from the Internet.

As the cost of empiricism falls, a cascade of empiricists of all kinds—economists, psychologists, political scientists—is flowing into universities and think tanks.[16] Thus, the declining relative cost of empiricism not only creates more empiricism but also orients our intellectual world toward the consequences of social policy, because these can now be better assessed. That is not to say theorists no longer play a role in creating social knowledge. We must have theories to test as well as data with which to test them. But empiricism also changes the way theorists operate: they give more consideration to whether and how their theories are testable, because a theory that is capable of verification is likely to have more staying power.

We see evidence of the rise of empiricism throughout the social sciences. Barry Eichengreen states that the last ten years have shown "a quiet revolution in economics."[17] He notes that high-quality empirical research is expanding dramatically and that "empirically oriented graduate students are the hot property when top doctoral programs seek to hire new faculty." He concludes that the twenty-first century will be "the heyday of empirical economics." Such empirical findings will improve our understanding of economic causes, perhaps permitting us to avoid financial crises of the kind the economics profession failed to predict in 2008.

The rise of empiricism has been the most important change in legal scholarship in the last decade. The Cornell Law School has created a new peer-reviewed journal focused on empiricism,[18] and a conference wholly devoted to empiricism and law is held every year. At the 2009 conference, 175 papers were presented, approximately 50 percent more than the previous year.[19] A workshop in empirical methods in law in the last decade has attracted five hundred law professors—approximately an eighth of all law research faculty—as students.

Disagreements about both empirical methods and results will not disappear, but the rewards of a culture of empiricism should make empiricists handle these disputes in a professional way, adjudicating them with reference to the best evidence and methodology.[20] As in other sciences, adherence to professional norms and interest in professional success will provide incentives for care and relative impartiality. Already social norms are evolving to encourage the disclosure of data so that other investigators can not only replicate results but also subject others' data to their own statistical tests.[21]

Yet another advantage of empiricism is to remind the political world that social behavior can have multiple causes. It may be that both merit pay for teachers—generally supported by Republicans and criticized by Democrats—and more spending for school instruction—generally supported by Democrats and criticized by Republicans—can improve a state's educational performance. As Marcus Aurelius is said to have observed: "We are too much accustomed to attribute to a single cause that which is the product of several, and the majority of our controversies come from that."

## Empiricism—Its Past and Its Future

Empirical findings have already made a difference to social policy in a variety of ways. Many of the most recent examples have been in the area of health and safety, because, as two commentators have noted, in that area there is a "particular public intolerance for policy failures."[22] For instance, graduated driver licensing—a practice by which young people are first issued restricted licenses and then permitted full driving privileges after a trial period—gained traction because of empirical findings. After empirical evaluations of successful use of such licenses abroad, they became widespread in the United States.

Another recent example shows that empirical studies can make a decisive difference to default rules in the financial area. Most 401(k) retirement plans had required employees to opt into their provisions. A paper then showed that an opt-out structure for such retirement plans rather than an opt-in provision was more likely to encourage employees to engage in 401(k) saving, meaning the opt-out plan would likely increase retirement savings and security.[23] Because of this analysis, the Labor Department provided a safe harbor for companies to offer opt-out plans.[24]

These are instances where empirical research was clearly a but-for cause of policy change. The complexity of politics makes it difficult to pinpoint such causes in most cases of social decision making, but empirical findings played an important role in other changes in social policy. One example is the change in antitrust rules on retail price maintenance, the practice by which manufacturers put a floor on the price at which their distributors are allowed to sell their products. This practice was long held to be illegal. But the U.S. Supreme Court in 2005 moved to a rule of reason in which it would not condemn these agreements without actual evidence of harm in the specific case.[25] In support of its conclusion, the Court cited studies showing empirical evidence that these agreements aided consumers by encouraging better advertising and service of products. Further empirical evidence since the decision has confirmed the view that retail price maintenance can have substantial benefits to consumers.[26]

The growth of computational tools and the spread of empirical culture should make empirical studies even more influential in the near future. To be clear, for the most part empirical studies will not directly change the minds of citizens. People are not statisticians and have better things to do with their time. But experts themselves affect policy both by setting the agenda for democracy and by deciding on important details of policy.

Empirical work is increasingly likely to affect citizens' views indirectly. As I will discuss in the next two chapters, empirical studies can influence prediction markets and the public policy debate carried on in the dispersed media. These new phenomena may then actually change some people's minds. These other technologies also have the capacity to amplify the effect of good empirical work. Their filtering can help the enterprise of social fact finding to inject a better ratio of signal to noise into political discourse.

From a historical perspective, the information age's greater capacity for empirical fact finding helps us begin to apply Francis Bacon's famous program for understanding the natural world to the social world. As economic historian Joel Mokyr describes it, the Baconian program for the natural world had three components. First, it sought to accelerate the pace of scientific knowledge.[27] To qualify as knowledge, information had to be cumulative, contestable, and yet ultimately supported by a consensus among researchers. In other words, researchers had to benefit from previous work; there had to be criteria to determine which of competing hypotheses was better; and ideas had to be open to contradiction.[28] Second, the Baconian research program needed to be directed to places with a high chance of payoff for human happiness, such as medicine.[29] Finally, this knowledge had to be made transparent by fair access and clear categorization.[30]

The contemporary empirical program for the social sciences partakes of all three of Bacon's conditions. It tests explanations offered for social behavior. It is cumulative, contestable, and seeks consensus. It is focused on important issues of social policy that have a payoff in social welfare, from improving education to curbing crime. The Internet and other information technologies are improving access and categorization all the time.

The knowledge of the social world will never be as secure from revision as that of the physical world. People are not billiard balls, and conditions of the future may be subtly different in a way that social science cannot yet capture. Moreover, even as to the past, empirical social science can generally only tell us that hypothesis is likely to be true within some probability. Yet probabilities of policy effects still provide information that can guide policy to the likelihood of better results and thus advance human happiness. That all uncertainty cannot be removed from evaluating the consequences of social policy is no argument against reducing as much uncertainty as possible.

An important difference between the Baconian program for the physical world and the Baconian program for the social world is that the latter is more dependent on a role for government. Government rules can help generate more opportunity for real-world experiments and empirical analysis.

## Empiricism and Information-Eliciting Rules

Just as empiricists become more valuable in an empirical age, so do structures of government that elicit information for empirical study. Government programs that actually put policies to an empirical test also become more desirable because we have better mechanisms to conduct testing. Besides the instrumental advantages of information-eliciting rules, they also create a better political culture. If politics focuses on actual policy results, politicians cannot as easily exaggerate the benefits of their policies and ignore their costs. Moreover, as information-eliciting rules create more testing, citizens will become even more insistent that policies be tested to assure their value.

The empirical age makes three kind of information-eliciting rules more desirable. The first two track the two kinds of empiricism discussed above. The first category encourages decentralization. By permitting jurisdictions to adapt different polices, a political system creates information about the effects of different policies that can then be tested through regression analysis and similar methods. The second category encourages randomization of the application of different policies. Such field experiments yield data to assess the results of those policies.

The third category is the simplest: rules that make government data more available in the most transparent and useful form. Such rules advance empiricism by offering more material for testing and assessment. The display of data also makes government more transparent, providing relative empowerment to those who want to organize on behalf of encompassing interests, such as improving education.

Each branch of government has a role to play in creating information-eliciting rules. Congress can systematically require that legislation consider the virtues of decentralization and randomization, require that government data be made available, and provide funding for empirical studies. The president can require executive agencies to engage in experiments within the discretion that Congress permits and prevent those agencies from squelching different approaches at the state level. The judiciary can adopt a jurisprudence fostering social discovery, creating constitutional space for federalism and thus decentralization and making sure that information about the differing results of policies is permitted to circulate freely.

To some degree this movement toward creating more space for experimentation is already taking place, if only inchoately and intermittently. In educational initiatives like the No Left Child Left Behind Act,[31] Congress has encouraged public schools to evaluate the success of their programs based on scientific evidence.[32] In the related Education Sciences Reform Act of 2002,[33] Congress created a federal agency to produce data about and analysis of programs to improve educational outcomes,[34] specifically calling for randomized trials to test educational programs.[35] The executive branch, particularly under President Obama, vigorously pursued initiatives like Race to the Top that encourage states to innovate and evaluate those innovations. It has also put more government data online. The Supreme Court under Chief Justices William Rehnquist and John Roberts has modestly revived constitutional federalism and curbed the Court's own tendency to mint new federal constitutional rights to be imposed on the states,[36] thereby enabling evaluation of different approaches to social policy.

But these movements have been largely piecemeal and have focused on particular policies and rights. Creating institutional rules for eliciting information can help assure a more systematically experimental approach to policy.

## Decentralization

Of all the American governmental structures that create information-eliciting rules, the oldest is federalism and the decentralization that it promotes. If policies are permitted to differ among jurisdictions, the consequences of good and bad policies may be compared and analyzed more easily. Such careful comparisons can help make manifest the consequences of good and bad policies. For reasons discussed earlier, such comparisons face more problems in establishing causation than randomized experiments, but randomization is not always either politically possible or desirable.

It is true that the decision about the proper amount of decentralization depends on considerations beyond eliciting information. Federalism, for instance, has other advantages: decentralization creates a market for governance by allowing different jurisdictions to compete in order to attract people and investment.[37] It also permits different policies to reflect different preferences in different states. But federalism can have costs as well. When one jurisdiction is able to impose harms on another (what economists call "negative externalities"), a more centralized government spanning the different jurisdictions sometimes has the capacity to address the externalities better than multiple jurisdictions can.[38]

Thus, the optimal degree of centralization of government policy remains a judgment call, depending on how much harm that centralization can prevent and how much competition and satisfaction of diverse preferences that decentralization can permit. But the possibility of sustained empiricism adds an important weight on the decentralization side of the scale: decentralization facilitates the empirical investigation of the differing consequences of social policy. Moreover, empiricism can also help us strike a better balance between decentralization and centralization. For instance, empirical research has suggested that concerns about states competing to attract business with worse environmental policies are somewhat overblown, thus suggesting more room for state experimentation in the area.[39]

Far from being a relic of the past, federalism's virtues are reinforced by modern technology, because our computer age makes federalism a far more effective discovery machine. With the rise of empiricism, Justice Louis Brandeis's famous praise of states as "laboratories of democracy" becomes more than a metaphor.[40]

At one time the United States had a system of information-eliciting rules in the form of constitutional federalism, whereby the federal government's power was strictly limited by the enumerated powers of its Constitution. Whatever the advantages and disadvantages of that system, such constitutional constraints are unlikely to be revived in all of their former strength. A large body of precedent now provides the federal government almost plenary power over economic matters. Given that the modern federal government is built on that precedent, it is very unlikely to be completely reversed. Because of the decline of constitutional restraints on federal power over economic regulation, it will largely fall to the political branches to create space for decentralized experiments in the economic arena.

Both sides of the political spectrum have shown interest in reviving federalism. The political right has long supported federalism, given classical liberalism's suspicion of centralized power and the analogy between competition among the states and competition in private markets. But the political left has recently become more supportive as well, because federalism may help demonstrate successful government initiatives to a public that has grown skeptical of big government.[41] In early 2011 President Obama suggested that he would like to revise his signature health care reform by allowing the states to design their own plans so long as they meet the federal requirements for expanding coverage.[42]

The federal framework suggested by the president's comments requires coverage of everyone on the theory that uncovered individuals will create an externality by driving up everyone's health bills across the nation. But within this framework, states could be free to determine how to cover everyone in their own state. Vermont, for instance, could experiment with a single-payer plan, and Texas could try a voucher system with vigorous competition among insurers.

To be sure, the division between state and federal responsibilities in both health care and education is itself contestable. Experimentation might well be more vigorous and more complete if states, rather the federal government, had to raise funds for their initiatives in health care and had greater say in exactly what services should be covered. But however this debate is resolved, states should be left space to experiment with policy so long as it does not cause substantial spillovers.

Congress should assure that the advantages of experimental federalism are not overlooked. Thus, each House should require that its committees formally consider the information advantages of leaving states free to legislate before enacting federal law that lessens the scope for state experimentation.[43]

The president also has a role to play in expanding empirically oriented federalism by preventing federal agencies from preempting state law with federal regulations when such preemption unnecessarily reduces the opportunity for experiment.[44] When statutes permit the government to waive federal regulations at state request, the president should order agencies to systematically consider the advantages of experimental learning that will accrue from such waivers.

The Supreme Court can advance empiricism by following a jurisprudence of social discovery, preserving and enforcing the constitutional provisions that assure information flows from individuals, mediating institutions, and state and local governments. The First Amendment and federalism are two of the American Constitution's most distinctive features. If protected, they assure that decentralized and competitive structures test different policy proposals and subject the results of these tests to vigorous debate. While Congress and the president can create new frameworks to make use of the new opportunities afforded by empiricism, the Supreme Court can build on structures that are already within the Constitution in order to help information bubble up from below.

The Court should retain what constitutional federalism remains. The Court has recently held that Congress may not regulate noneconomic matters unless they are part of a larger economic scheme.[45] The pragmatic justification for this line of cases is that such noneconomic matters, like the question of whether guns may be carried near schools, tend not to impose substantial spillovers among the states. Here the Court can preserve a constitutional federalism that the federal government cannot displace. As a result of this jurisprudence, different states can adopt different policies such as whether to ban the carrying of guns near schools. Empiricists can then test whether such bans increase or decrease student safety.

The Supreme Court can also advance a jurisprudence of social discovery by declining to expand its own jurisprudence of substantive due process, the controversial doctrine by which it reads the due process clause to impose uniform federal substantive rights on the states.[46] Curbing this doctrine would give states the opportunity to experiment and compete in decisions about the optimal bundles of rights, making trade-offs between liberty and other concerns, including traditional morality. Technological acceleration provides an even stronger reason for creating such a space for social discovery, as the trade-offs may change more quickly, and empirical evaluation of the results of different choices will be even more useful.

Thus, for example, the Supreme Court should stay its hand on mandating a national right to same-sex marriage.[47] Social scientists will be able to study the actual effects of same-sex marriage by comparing jurisdictions with same-sex marriage and those without it. They can then better investigate the claims by proponents that same-sex marriage will help stabilize same-sex relationships[48] and the claims by opponents that same-sex marriage will mean couples generally will take their marriage obligations less seriously.[49]

Of course, many people on both sides of the debate may not care about these results, because they believe that bad consequences cannot impugn the status of same-sex marriage as a universal right or that good consequences cannot ameliorate its natural wrongness. But some people in the middle do not share such deontological absolutism. Proponents and opponents of same-sex marriage recognize this fact; that is why they make arguments based on consequences.

As technology accelerates, new social questions will be framed in terms of rights, like the right to make a designer baby with genetic enhancement or the right to clone. Because these questions raise uncharted trade-offs between liberty and other social norms, they are better addressed through the feedback about their consequences that federalism affords. Just as universities hire more empiricists than theorists in an age amenable to empiricism, so the Court should create a more empirically oriented jurisprudence rather than rely on its intuitions about the proper set of substantive rights.

In contrast, from an informational perspective there is less objection to judicial activism from state courts interpreting their own constitutions. When the Massachusetts Supreme Court legalized same-sex marriage, it permitted the evaluation of a new policy. The decision provided two different kinds of information. Besides permitting an assessment of the results of same-sex marriage, it permits the assessment of the results of different kinds of political regimes, because Massachusetts is a state that entrusts more responsibility for making social policy to the judiciary than most other states.

The Supreme Court also creates better information-eliciting rules when it declines to expand the application of the Bill of Rights against the states. Strong textual and structural reasons already militate against interpreting the Fourteenth Amendment to apply the Bill of Rights against the states with the full rigor that they apply against the federal government.[50] But the greater power of empiricism adds a pragmatic reason for giving states a more substantial margin of appreciation in deciding how these rights should be deployed, because their different decisions will help evaluate the consequences of different scopes for rights.[51]

The Court has not yet offered the promotion of empiricism as a rationale for permitting state experiments to go forward despite claims that they violate the Bill of Rights. But it has been frequently remarked that the Roberts and Rehnquist Courts have enforced the Bill of Rights less strictly against the states, thus effectively moving toward a jurisprudence of social discovery in this area as well.[52] In the most important education case in decades, the Court in *Zelman v. Simmons-Harris* upheld a local system of school vouchers against a federal Establishment Clause challenge.[53] This holding permits mediating institutions like religious institutions and private schools to compete with the government on more equal footing in order to help test what educational structure will deliver the best results.

Indeed, this decision illustrates that permitting experimentation among the states in one area facilitates a more general culture of experimentation and democratic updating. The Court's decision in *Zelman* makes it far easier to test whether vouchers can help improve performance in our education system.[54] And if vouchers do improve performance, schools will produce citizens who are better able to process information and make wiser decisions, thus rendering democracy as a whole a more effective social discovery machine.

While liberals frequently resort to the Establishment Clause to trump state laws, the same analysis should apply to provisions of the Bill of Rights to which conservatives appeal. The Supreme Court has held that the right to keep and bear arms applies to the states,[55] but the Court should not apply this right with the same vigor against the states as it does against the federal government. Greater scope for experimentation with gun control will allow better evaluation of the appropriate scope of gun rights no less than a relaxed interpretation of the Establishment Clause will allow for state experiments on the role of religion in civic life.

Of course, at some point there may be sufficient results from experimentation in the states to determine that there is a need for a strong and uniform federal right. However, the constitutional amendment process is the better way to make that decision than leaving it up to the intuition of Supreme Court justices. Technology is changing the social landscape of society so quickly that we need to be very certain about the beneficence of a right before entrenching it nationally for the indefinite future. The constitutional amendment process helps provide this certainty by requiring a relatively stringent supermajority for adding provisions to the Constitution, mandating substantial societal consensus.[56]

Federalism is a particular instance of a more general social policy—decentralization—that favors experimentation. Just as federal legislators should allow space for state experimentation, state legislators should devolve responsibilities to those smaller localities when those localities are better positioned to test different policies. Localities also should consider the possibilities of experimentation in the policy structures they create. For instance, because charter schools enjoy greater independence to set educational policy, they create more opportunities to test what ideas work best.

### Randomizing Policy

The government can also create many more field experiments, following the lead of private firms. Businesses have been creating field experiments to make decisions on an ever broader scale.[57] In one example, Google experimented with different colors for its search links in order to determine what color those links should be.[58] Banks and other corporations have used randomized trials to improve marketing; they have reduced costs by doing more of these trials on smaller scales.[59]

The federal government has already engaged in some random social experiments, but they have mostly centered on the disadvantaged, such as welfare recipients.[60] Thus, various studies have randomly assigned recipients of government benefits to different treatments. Some have focused on how differences in job training or counseling in government programs affect employment.[61] There is room for more such experiments. The extent to which the prolonging of unemployment benefits results in longer unemployment is still controversial and could be tested through randomization.[62]

However, the government has almost never conducted randomized tests of policies affecting middle- or upper-class individuals, corporations, or the structure of government itself. Policies affecting individuals might include the effect of electricity charges that vary with time or how differing deductibles for health care coverage would affect medical usage and health outcomes.[63] The efficacy of school vouchers is one of the most important, yet contentious issues in education today and can best be assessed through randomized testing.[64]

Randomization can also be applied to corporate regulation. There remains a substantial debate over whether the conditions the Federal Communication Commission imposes on public broadcasters are effective at meeting their purported objectives, like encouraging diversity of programming. The FCC could randomize some of the policy conditions to test their effectiveness. Randomization can also be applied to government procurement, testing the wide variety of auctions government can use for its contracts.[65]

We must recognize that field experiments will not routinely provide conclusive proof of what policies to pursue, because social phenomena often involve complex interactions.[66] Even if pupils in Minneapolis assigned to charter schools performed better than those assigned to public schools, it might turn out that the cause of their superior performance lies in factors that are peculiar to Minnesota. Nevertheless, such an experiment has value, because it will change our view about which policy should be pursued locally. Unless Minnesota is unique, the result will probably change our estimate of how effective vouchers are likely to be elsewhere. Empirical knowledge gained by decentralization or randomization will become most powerful only by accumulation, but that is all the more reason for Congress to create institutions that gather information systematically.

Thus, the salient question in randomization is whether this additional information is worth the cost of the experiments and, if so, how to best design a policy to generate it. Congress can create an office modeled on the Congressional Budget Office with the authority to recommend the insertion of specific provisions, including funding, into legislation to evaluate the consequences of the policies that the same legislation puts in place.[67] The office could also become a clearinghouse of knowledge, transporting to the public sector some of the ideas of randomization that have been successful in the private sector. Such a new institutional structure also signals more generally to the bureaucracy that it should emphasize experimental methods at the expense of more top-down methods of social ordering.

It might be objected that it is wrong to apply the law differentially to individuals for social gain. But in medicine randomized trials of new pharmaceuticals are routinely undertaken.[68] There, as here, the justification is that we are not confident about which particular course of action will provide more benefits. Randomization of social policy should be used only where there is a genuine controversy over which social policy to follow. Fortunately, those are the general circumstances in which randomization may enjoy political support.[69]

It is true that patients must voluntarily agree to medical trials while people would not normally have the option of avoiding the randomized regulation. But that difference results from the structure of preexisting legal rights. Patients have the right to refuse treatment and thus cannot be forced to participate in a medical study. People do not have the right to refuse to follow a valid regulation and hence can be forced to obey even a regulation that has been chosen at random, so long as those regulations are otherwise legal. Even without randomization, citizens are essentially used as experimental subjects whenever politicians impose regulations or taxes in the hope of social improvement. Policy randomization can put such policy ideas to a more rigorous test, increasing public knowledge and helping to end those programs that are counterproductive. Randomization thus has a powerful justification whenever the policies to be evaluated have plausible justifications themselves.

Both executive and judicial precedents suggest that randomization is legal. Most dramatically, the government instituted a lottery to determine who would be drafted in the Vietnam War.[70] While this randomization was not designed for experimental reasons, it had far more dramatic results than the kind of randomized policies recommended here. Some young men faced a much higher risk of death than others as a result of a random drawing. And this method of choice did not allocate that risk on the basis of willingness or capacity to serve.

Although the Supreme Court has never reviewed a challenge to randomization, the federal court of appeals in New York rejected the only substantial legal challenge to randomization within regulatory policy for reasons of social experimentation.[71] In that case, New York State required family members in the households of public recipients to engage in training or working but imposed these requirements only in certain districts, choosing them on a random basis.[72] The court responded to the equal protection challenge by holding that the policy of social experimentation was wholly rational, given the importance of experiments to social policy.[73]

## Data

Data provide the foundation for empiricism. Over the years, government has become more transparent; the federal Freedom of Information Act was landmark legislation in this regard.[74] But the rise of empiricism provides a stronger rationale for making public everything that government does (outside of sensitive national security matters, business trade secrets, and matters that trench on personal privacy). Thus, government data should be posted automatically and in machine-readable form so that it can be easily used for empirical research. As Beth Noveck has suggested, third parties should also be invited to make the data more understandable through the use of charts, graphs, and videos.[75]

The process of making government data accessible will be a continuing enterprise, because technological acceleration will improve our existing mechanisms of distribution, such as the World Wide Web. For instance, the so-called semantic Web will imbed information in Web pages so that machines can recombine and present information with less human intervention.[76] The government needs to assure that its data will be in a form that comports with such improvements.

President Obama has started to make data more available through his Open Government Initiative.[77] It creates a presumption of transparency for data and encourages data to be published online.[78] A website (www .data.gov) has already been established to be a kind of clearinghouse of government data.[79]

Unfortunately, Congress eliminated much of the funding for this initiative in a 2011 budget deal; because of concern over the deficit, funding was cut from $35 million to $8 million.[80] This reduction is extremely shortsighted. Online data will make it easier to propose efficient revisions to government programs, saving money in the long term. Greater transparency makes programs that benefit special interests more apparent to the public, mobilizing encompassing interests against these exactions.

The judiciary needs to do a better job of getting its data online. PACER, the electronic mechanism that provides access to federal court records, needs to be updated; the current system that it runs has been well criticized as antiquated and excessively expensive.[81] Information contained in public court filings must be put onto the Web in its most usable form. Use of the system should be free, at least for research purposes. Of course, such electronic databases raise issues of personal privacy, such as the need to avoid disclosure of Social Security numbers and addresses, but technology itself can aid in creating automatic programs to make such redactions.

Because of the benefits of comparing state policies, empirical investigation of policy would benefit from similar initiatives at the state level. Given that each state can benefit from knowledge about other states' policy initiatives, data about the programs of individual states is a good that redounds to the benefit of the whole nation. Thus, the federal government should subsidize these state initiatives in the national interest.

Finally, the government should combat reporting bias in empirical studies. Reporting bias flows from the failure of researchers to make public studies that do not have statistically significant results or otherwise have flaws.[82] The practice of underreporting may make these studies that are reported unrepresentative. Thus surveys of the actual evidence in the field, like those that perform meta-analyses of all studies, may be distorted.[83] Yet each individual researcher has little incentive to report studies that do not reach statistically significant results.

As a condition of receiving funding for a study or using a study to get regulatory approval, the government should require that any studies performed by the recipient or applicant be reported whatever their results. For instance, obtaining approval for new pharmaceuticals from the FDA should require companies to commit to making public all of their studies, whatever their results or lack of results. Another advantage of such mandated disclosure is that it may prompt professional associations to adopt ethical rules requiring fuller disclosure of data and reports from their researchers.

Most of the information-eliciting rules recommended here will require some modest government funding. But just as public funding for natural science is useful when it can pay off in research that benefits the nation, so is funding for social science research. The computational revolution has made such spending more beneficial. Information-eliciting rules should become part of society's software of social norms that take maximum advantage of the improvements in hardware.

### Changing Laws to Promote Private Incentives for Research

These large structural changes to government are the most important way we can reorient political culture to a greater concern with policy conse-quences and a greater capacity to evaluate them. But laws that provide incentives for private individuals and businesses to offer data and analy-sis to the public also become more valuable with the rise of empiricism.

Thus, the political branches should vet laws and regulations to en-courage the creation and structuring of more data. At the time of this writing, Google is trying to enter into an agreement with publishers to create an online library where all books can be searched. In analyzing this proposed settlement, executive agencies should give substantial weight to the potential of this new digital catalog for data extraction, not only by humans but by machines.[84] Such data mining, particularly when linked to other available data banks, may enable us to help evaluate past policy effects on subtle matters like cultural change.

We must also be very careful that rules for reviewing research do not become so onerous as to reduce the number of experiments that ana-lysts can undertake. In particular, Institutional Review Boards (IRBs) mandated to oversee government-supported research at universities have been severely criticized as imposing substantial costs on research without any empirical evidence that they are providing benefits.[85] The govern-ment should carefully review the rules that govern the operations of these boards. One solution would be to create a more permissive regime, in which different universities could experiment with different structures and levels of scrutiny. The results could then be evaluated.

### Information-Eliciting Rules and Political Culture

Beyond their instrumental advantages, information-eliciting rules help create a better political culture. First, the culture becomes more self-consciously experimental. Citizens expect to hear about the results of policy when government is structured to evaluate policy results. The claims of politicians and interest groups then are more readily tested against the consequences of regulation and legislation. Such a culture will not be created overnight, but over time it can reorient political discourse from grandiloquent claims of social transformation toward a greater focus on what actually works.

Second, these rules encourage a greater humility in politics. Anyone involved in private markets recognizes that their ideas for a new product, however good in theory, must be tested by the market. Similarly, information-eliciting rules subject government policies to more intensive scrutiny and testing, forcing politicians and bureaucrats to see whether their ideas succeed or fail.

Finally, such rules help us concentrate on what we have in common—the facts of the world—rather than the ungrounded intuitions or personal circumstances that may divide us. Thus, information-eliciting rules can cultivate the best of the Western Enlightenment tradition: a willingness to change our opinions as the facts change. That openness to evidence is an important component of what Judge Learned Hand called "the spirit of liberty which is not too sure that it is right."

# CHAPTER FOUR

# Unleashing Prediction Markets

POLITICAL PREDICTION MARKETS—markets that allow the public to speculate on election and policy outcomes—have the potential to improve the capacity of democracy to update on the facts in our day, just as the rise of the press improved its capacity in an earlier era.[1] These markets can elicit information about the likely effects of policies even before they are implemented from those who are most knowledgeable about their effects.

Prediction markets temper three of the largest problems of politics. First, they offer a mechanism for overcoming what has been called the "basic public action problem" of collective decision making—the difficulty of persuading individual citizens to provide information that is useful to the whole community. This problem goes back to the origins of democracy: "How does Athens know what Athenians know?"[2] Prediction markets give citizens stronger reputational and monetary incentives to inject their private information into the public domain, thus improving social knowledge. Second, the markets help mix expert and dispersed opinion. Expert views influence prediction markets, but they can be tested by outliers who are willing to put their money where their mouth is. Finally, prediction markets are able to draw together information into a single place and numerical form, economizing on the attention of citizens. This latter capacity is particularly important today, because in an age of accelerating technology, people are likely to be distracted from the business of public life by the more compelling entertainment of private life.

Prediction markets can enrich democratic deliberation only if our legal policy toward them is changed. Currently our law inhibits their growth. For the future, the government must not only unchain these shackles but also fund experiments to help discover the most promising designs of such markets.

## The Nature of Prediction Markets

In prediction markets traders can buy shares that pay off if a specified event occurs, like a candidate's victory in a presidential election.[3] Some of these markets are already in operation. An Irish prediction market, Intrade, for example, currently has created a market for predicting the presidential election in 2012. Assume, for instance, that on November 30, 2011, on Intrade one could have bought for approximately $5.80 a share that paid $10.00 for a victory by the Democratic presidential candidate in 2012 and for approximately $4.10 a share that will pay $10.00 for a victory by the Republican presidential candidate.[4] These shares can then be traded until the event to be predicted occurs—in this case the result of the presidential election.

As of November 30, 2011, such a market not only predicts a Democratic victory in 2012 but also provides an estimate of the likelihood of the victory, given the information available at that time. It has been debated in the technical literature on prediction markets whether such prices translate into probabilities. The consensus answer, based on both models and empirical comparison of predictions and results, is that they do.[5] The markets designed to predict the vote shares of candidates have generally proved more accurate than polls.[6]

In the example described above of a market designed to predict the chances of a presidential victory, if the probability of a Democratic victory were higher than 58 percent on November 30, 2011, individuals would likely make money by buying shares in a Democratic victory. If the probability were lower, individuals would have an incentive to sell them, depressing the price. The market also gives a rough probability of a judgment that is not likely to happen at that time—a Republican victory.[7] Thus, it is important to note that the event with the highest probability in a market will not invariably occur. A prediction market is accurate if the events happen in proportion to the predicted percentages. Thus, underdogs at $4.00 a share on Intrade should win approximately 40 percent of the time.

Gambling on events is itself not dependent on technological acceleration; bookmakers made odds on the election of medieval popes.[8] But just as the influence of the written word transformed social governance with the invention of the printing press, so prediction markets can alter social governance with the invention of the Internet. The Internet permits these markets to operate on a much larger scale and with far lower transaction costs than in the past.[9]

It is thus now possible to improve information about public policy by creating markets for predicting thousands of events, including those that are conditional.[10] A conditional event market permits people to bet on an event that is conditional upon another event occurring. Conditional markets are important for consequentialist democracy, because they help predict the consequences of an election or a piece of legislation *before* the votes in the election or on the legislation have been cast.

Commercial gambling sites have already made markets in conditional events. Before the 2008 election, on the market Intrade one could have bet on the economic growth rate that was conditional on Senator Barack Obama being elected and conditional on Senator John McCain being elected.[11] Such markets have the potential to help evaluate candidates,[12] because they can predict the candidates' performance on various dimensions of social well-being.

Note that conditional markets can be run in a way that does not require money to be returned if the condition is not met. The market simply requires a bet on all the conditions. Thus, in a two-candidate race, a bettor predicts the particular economic growth rate that will result upon the election of either candidate. Given that one of the candidates will be elected, the market maker can determine whether the bet will be paid off. The greater impediment to the flourishing of prediction markets is that a bettor loses the time value of money for conditional bets whose payoffs can be determined only far in the future. As discussed below, government can temper this problem by experimenting with providing some modest subsidies to markets.

Conditional markets can also be made on the consequences of specific policies as opposed to electoral results. For example, one could bet on the likely effect of a change in tax policy. For instance, in 2011 President Obama proposed a payroll tax holiday for a year. Prediction markets could be made on the effects on economic growth in 2012 that are conditional on the passage of his tax proposal and conditional on its failure. The growth rates would be measured by a certain metric, such as the official government growth rates. Traders would bet on a variety of bands of economic growth. The amount could climb from 1 to 4 percent in increments of one-tenth of a percent. One would thus learn the prediction market's estimate that economic growth would attain a certain level conditional on the normal payroll tax rate and conditional on a certain payroll tax cut.

Because we are interested in longer-term effects on economic growth, similar conditional markets could be made for the economic growth rates in 2013 and 2014 and on the unemployment rates. Since the purpose of the payroll tax cut is to substantially increase economic growth and employment, such markets would be useful, helping us understand the degree to which the proposed policy increased economic growth and employment from year to year. For instance, by expanding the years in which income growth is predicted, we could evaluate whether a payroll tax cut simply shifts income growth from future years to the current one.

Markets predicting variable consequences of an event have already been made. Intrade generated markets predicting the number of seats a party would gain in the House of Representatives after the 2010 election.

It offered shares in possible net gains of seats—none, more than five, more than ten, and so on. Of course, the price tended to decline for shares with greater and greater gains of seats. But the market as a whole provided a distribution of probabilities of gains by a party of seats in the House.

Hundreds or thousands of conditional markets can be created in public policy. They could include markets on the economic effects not only of specific kinds of tax cuts but also of specific government spending. Moreover, the markets could focus on a variety of consequences of public concern, from economic growth to economic inequality, as assessed by objective measures like the GINI coefficient, which measures such inequality.

To be clear, such conditional markets will not always settle policy arguments because of the problem of teasing correlation apart from causation.[13] For instance, assume that the market predicts higher growth if no payroll tax cut is made than if the payroll tax cut occurs. Arguably, some social phenomenon correlated with the failure of the payroll tax cut could be responsible for the additional economic growth. Yet the prediction market is still valuable, because it forces the articulation of other factors that may themselves be relevant in setting policy or in voting. And unless it is plausible that these factors are very closely correlated with economic growth, the market still helps show the degree to which a specific payroll tax cut makes a contribution to economic growth.

Another answer to the problem of correlations is to create even more prediction markets.[14] Assume the markets suggest that a world with the specific payroll tax cut will not raise growth in 2012 as opposed to a world going without the tax cut. A decline in President Obama's political power is one occurrence that might be plausibly correlated with the failure of a payroll tax cut. This decline could raise the predicted level of economic growth conditional on the failure of the payroll tax, because investors would take its failure as a signal of Obama's likely replacement by a more business-friendly Republican president in 2012. Conversely, the success of a payroll tax cut is more likely to suggest that the president retains enough popularity to be reelected.

To gauge the extent of the correlation between President Obama's popularity and predictions of economic growth, prediction markets could be made with two conditions. For instance, one market could be made conditional on both the failure of a payroll tax cut and an Obama *victory* in 2012, and the other market could be made conditional on the failure of a payroll tax cut and an Obama *defeat*. The differing results would help make clear whether the payroll tax cut had an effect independent of the perception of Obama's political strength.

As Professors Todd Henderson, Justin Wolfers, and Eric Zitzewitz have observed, empiricists can also try to tease causation from correlation by conducting event studies on prediction market data.[15] Event studies follow the prices of an item over time and try to assess the effects of one kind of event on the price. An event study could follow prices in a market that was conditional on the presence of payroll tax cut, investigating the relation between that price and the standing of President Obama in the polls. If the market tended to predict higher economic growth whenever the president went down in the polls, such an event study would suggest that it was the prospect of an Obama defeat that was driving the prediction of higher economic growth rather than the payroll tax cut. But if Obama's standing had relatively little effect, the event study would suggest that the president's election prospects and the effect of a payroll tax cut were not highly correlated.

These event studies are not simply speculative. One paper looking at the relation of the prediction market for the presidential election in 2004 and the stock market compared the movement of prices in these markets.[16] Relying on this analysis together with other data, the authors concluded that equities would become more valuable under a Bush presidency than under a Kerry presidency.

The consequences on which prediction markets focus can be anything that people might think is relevant to public policy. For instance, the effects of President Obama's health care plan were a matter of intense debate in 2010. Before the passage of that plan, prediction markets could have been made on the various health care cost measures, such as Medicare expenditures and total health care expenditures, conditional both on the plan's passage and on its defeat. Ultimately health care policy affects the rate of medical innovation, particularly in an age of technological acceleration. Thus, markets could be made on death rates, conditional again both on the passage of the health care plan and on its defeat.

Markets can also be created on the effects of accelerating technologies themselves. Some of these technologies might be very valuable, but some are potentially harmful. Thus, some of the tiny particles made in the process of nanofabrication may cause deadly pollution because our own bodies are not adapted to dealing with these foreign bodies. A prediction market can be made on the whether a specified set of particles will have harmful effects according to some specified metric. Similarly, prediction markets can be based on some of the alleged dangers of genetically modified foods—for example, whether allergy rates increase with their introduction. Predictions about future technologies can also inform policies today. If solar energy will be cost-effective in ten years, the government may invest less in other alternative energies and regulate greenhouse gases less severely than if carbon-based fuels were to continue to dominate energy production. Thus, it may be helpful to have prediction markets that attempt to gauge the future price per kilowatt of solar power. Prediction markets may then prove useful for collective decision making about both traditional issues, such as what are the best policies for economic growth, and issues created by accelerating technology itself, such as what are the possible costs and benefits of regulating nanotechnology.

## The Advantages of Prediction Markets

The greatest advantage of prediction markets over other sources of information is that they aggregate information from those who have incentives to get their predictions right. If you do not know anything about the subject in which shares are being sold, it is a bad idea to participate, because you are likely to lose out to those who do.[17] Moreover, like the stock market, a prediction market elicits information from those who would otherwise be unlikely to offer it to the public, because the market provides them a monetary incentive to act on their private information.[18] Prediction markets also instantiate the economist Friedrich von Hayek's great insight: information about the world is dispersed among many people.[19]

More specifically, prediction markets increase the common social knowledge that makes for better policy and a more stable society. First, prediction markets and empiricism feed off one another. Prediction markets take account of empirical data and add information from other sources to make better predictions than could be based on empiricism alone. Empiricism tells a backward-looking story of social life;[20] prediction markets help convert that information into a forward-looking perspective. Prediction markets thus make empiricism more valuable because, as Søren Kierkegaard reminded us, life is understood backward but lived forward. Of course, Kierkegaard was hardly a social scientist, but the point can be stated in terms of social science: the future is not in the sample of data that empiricists have analyzed in the past. Information markets thus help make out-of-sample predictions.

This capacity shows why some criticisms of the accuracy of prediction markets based on regression analysis of these markets miss the mark. Two political scientists have argued that someone could take poll data and add information about what had happened in a set of previous elections and use a statistics-based regression analysis to predict elections better than prediction markets.[21] But no one knows whether these scientists would have chosen this particular data set and constructed this particular model to predict elections before they knew the results of the elections.[22] Moreover, in the future people would debate whether their regression model is still accurate, given changed circumstances. But this study does underscore another virtue of prediction markets: if markets elicit more empirical work, this empirical work should inform future prediction markets.

Prediction markets also help empiricism more generally, because they generate more data on which empiricists can perform event studies. Thus, by predicting the probability of events that have not yet happened, prediction markets create possible worlds, multiplying the information available for understanding our actual world.

A second advantage of prediction markets is that they combine expert and nonexpert opinion, because nonexperts can bet against expert results. Some expert studies may be distorted by the ideological views of social scientists that are unrepresentative. Prediction markets create a market check against that bias: the public can bet against the consensus of experts.[23]

The power of market tests was demonstrated in the 1970s when Paul Ehrlich, one of the leading experts on population growth, argued that population was outstripping natural resources. In 1980 economist Julian Simon offered and won a bet that key natural resources like tin would decline in price by 1990, suggesting that fears of coming scarcity were overblown.[24] Prediction markets would institutionalize such challenges to expertise.

We can imagine prediction markets being developed to predict future temperatures in order to test some of the claims of climate change. It is true that some future effects of climate change may depend on policies adopted, and thus conditional markets may be useful in evaluating the predictions of various climate scientists. But some climate scientists predict significant warming in the relatively near term, regardless of the adoption of any plausible policy option. Passing this test would help in putting to rest attacks that the claims of climate scientists are infused with political bias. Prediction markets may also help address the degree to which particular kinds of tax cuts pay for themselves by increasing economic growth. Prediction markets can be made on the accrual of revenues conditional on a tax cut and conditional on the lack of a tax cut.

A third advantage of prediction markets is that they focus information about likely policy effects and make evaluation of this information easier. Citizens have trouble making sense of disparate information about politics.[25] In particular, they have difficulty analyzing statistics, discounting flimsy data, and giving proper weight to strong data.[26] Prediction markets integrate the data for the public.

Better focus is even more important in an age of technological acceleration. The very richness of information generated in our world creates the danger that even public-minded citizens will drown in data. As Herbert Simon remarked, "A wealth of information creates a poverty of attention."[27] Prediction markets take all of the information and encapsulate it into numbers that are relatively easy to understand.

Political scientists have likened voters' search for political knowledge to a drunkard's search for his keys.[28] They look where there is light, regardless of whether that is the best place to search. Thus, since the mainstream media often focus on the personal characteristics of candidates, voters try to extrapolate from personal characteristics to policy positions and effects. In contrast, prediction markets on policy shine light on the likely effects of public policy. The markets thus would help people make the key connection that is central to democratic updating: the relation between politicians' policy proposals and the proposals' effects on the future of the nation.[29]

## Responses to Critiques of Prediction Markets

Prediction markets have been criticized on a variety of grounds. Before reviewing these criticisms, it is important to remember the nirvana fallacy: the mistake of comparing an institution as it actually operates to some unrealizable alternative. Prediction markets may have flaws, but so do other ways of generating accurate information about public policy. The question is not whether markets are perfect but whether they are better than other mechanisms of policy assessment, either on their own or in conjunction with other institutions.

For instance, prediction markets are correctly said to have limited utility in predicting matters like the choice of a nominee to the U.S. Supreme Court when that decision is to be influenced by only a few individuals.[30] It is also said that information markets are not good at predicting idiosyncratic matters, such as trial results in particular cases.[31] But there are few other ways that are better; decisions that depend on the psyche of a few individuals are inherently hard to predict. Moreover, events subject to the vagaries of a few decision makers are not the most important issues about which democracy needs predictions. Important phenomena such as economic growth or unemployment are affected by the future decisions of millions of people.

The sometimes rapid gyration of election markets on Election Day has also occasioned adverse comment.[32] But at times an outcome in a close election is uncertain and new information changes the probabilities of victory very rapidly.[33] There is little evidence that non-market mechanisms are better at predicting in such circumstances, so long as the market has enough traders and liquidity.

Four more substantial criticisms have been made of markets. The first is that prediction markets do not actually add predictive value. For instance, it is claimed that prediction markets in elections just move with the latest poll results.[34] Even if they do better than a poll, the argument runs, their superiority arises from their ability to aggregate information, not to offer new information. Even if this critique were accurate, public policy would benefit from the aggregative effects of prediction markets. All polls are not equally accurate, and the market would weight them according to their credibility.[35] Prediction markets, then, focus the attention of voters on the single best indicator of likely election results rather than leaving them to wonder which poll is credible.

But this critique is not true of election markets. Election markets are generally more stable than polls in the run-up to an election. Polls go up and down depending on the events of the day. But participants in the market bring knowledge of which events are likely to have lasting effect on public opinion. An excellent example is the effect of the killing of Osama bin Laden on the likelihood of President Obama's reelection. Obama's poll numbers went up dramatically for some weeks, but after a quick initial spike on Intrade, his chances of reelection increased much more marginally. Experienced election observers recognize that most of the political advantage from that foreign policy victory will have worn off by Election Day. Another example came in the 2011–2012 Republican primary season. After turning in a series of strong performances in the candidates' debates, Newt Gingrich, the former Republican Speaker of the House, vaulted to the head of the polls in Iowa and in the nation as whole. Many in the press labeled him the front-runner, but he always remained behind Mitt Romney in the prediction markets.[36] Despite the polls, markets recognized Gingrich's vulnerabilities and Romney's financial capacity to remind voters of them.

Moreover, this critique is not applicable to subjects for which there is no prepackaged information, like polls. Players in a conditional prediction market, like that of the effect of a stimulus package on economic growth, may have empirical knowledge of the effects of past government stimulus spending, but the relevance of these studies depends on the state of the current world. This market gathers dispersed information about this state, weights its relevance against past studies, and consolidates it into a prediction. This process creates new and valuable information.

A second critique is that, like other markets, prediction markets are subject to behavioral imperfections, the most notable being *long-shot bias*—the tendency of bettors to overweight low-probability outcomes—and *bubbles*—the tendency of markets to overshoot the mark because of expectations that prices will continue moving in the direction they have before. Were these criticisms to be completely true, it does not follow that prediction markets would be useless. Even with inaccuracies, prediction markets are likely better than other mechanisms of predicting policy consequences, or at least better in conjunction with other methods. For instance, if prediction markets have a long-shot bias, there is some evidence that they may still correct for even greater biases people have in estimating the probability of long-shot events in the absence of markets.[37]

In any event, problems created by these market imperfections are exaggerated. It is true that there is a long-shot bias in the betting market.[38] The bias is probably an example of a heuristic bias; most people are not good at discerning the difference between small and very small likelihoods.[39] But even if there is a long-shot bias for low-probability events, prediction markets can be useful in assessing the likelihood of higher probability events. Moreover, predictions for long shots can be seen through the prism of the bias and then appropriately discounted. Finally, while biases exist in markets dominated by amateurs with high transaction costs, they are not as prevalent in markets dominated by professionals with lower transaction costs.[40] As prediction markets gain traction, particularly with government support, they are likely to attract more professionals.

A closely related critique is that prediction markets are not helpful in assessing the probability of unlikely catastrophic events. The first element of this criticism—the problem of long-shot bias—has already been addressed. It may even have less bite in this context. One important issue for long-term catastrophic events is determining the most likely of unlikely events, because we have limited resources to guard against catastrophe. Since presumably long-shot bias applies to all such unlikely events, prediction markets may help us decide where to concentrate resources. The second component of this critique is that the market may itself influence the likelihood of action to avert the catastrophe, thus distorting the results.[41] But this critique misses the power of conditional markets. One could run a market that is conditional on the best strategy to avoid a catastrophe and conditional on nothing being done. One caveat: the catastrophe must not be so cataclysmic as to significantly affect the chance of the prediction market making payouts.

The literature supporting the existence of bubbles comes less from the betting markets than from the stock market. Bubbles are also said to stem from a heuristic bias—the representativeness bias—which makes people tend to think the future is going to be like the past.[42] Thus, if stock prices go up, there is a tendency to believe that the climb will continue. But the existence of bubbles is disputed. Some economists believe that the most famous examples of supposed stock market bubbles, such as the crash of 1987 or the Internet bubble of the late 1990s, can be explained by changing fundamentals in the real world, such as interest rates.[43] Thus, in markets with large volumes like the stock market, it is not even clear that bubbles exist, let alone that their existence suggests there is some other mechanism better than markets for assessing value. Moreover, there have been few patterns that resemble long-term bubbles in prediction markets yet. A candidate's chances of victory occasionally soar, but these spikes have generally proved evanescent. Such irrational movements are more likely the result of thinly traded markets.[44] Finally, the nirvana fallacy remains still relevant. Prediction markets, like the stock market itself, may still yield valuable information, even if on occasion one floats on a bubble.

A government fiasco prompted another concern about prediction markets. In 2003 the Defense Advanced Research Projects Agency (DARPA) conducted a private prediction market among analysts to predict foreign policy events. In comments on its website, one kind of event mentioned was an assassination. Political representatives and others seized on this stray comment to assert that it is immoral for people to profit from terrorism.[45] This criticism was unjustified, however, because the purpose was to generate information to *prevent* terrorist attacks. If someone invents a new surveillance mechanism that spots terrorists, his income would also derive from the possibility of terrorism, and yet we would not consider such an invention immoral.

This kind of market might become morally problematic if it were open to the public, as most effective prediction markets will be. The fear is that terrorists or those tipped off by terrorists could bet on terrorist events, be proved right, collect the money, and then use the proceeds to fund more terrorism. Even this fear may well be overblown. Betting in such a market might well enable the government to track down the terrorists. Moreover, other markets may offer even larger possibilities of gain with less possibility of detection. One could have made an enormous amount of money by shorting global stock markets on September 10, 2001. Nevertheless, the perception that terrorists could benefit from a market devoted to predictions related to terrorism would be intolerable to the public legitimacy of such markets. Thus, prediction markets should be prohibited when they may provide incentives to people to perform illegal acts. But such concerns do not undermine most uses of these markets.

The fourth and most serious critique of prediction markets stems from a concern about manipulation.[46] It is generally thought that this risk is not substantial for long-term prices in most financial markets. Traders have incentive to buy or sell shares to get back to the level justified by information if manipulators bid up or sell shares without any informational basis.[47] But given that the focus here is on using prediction markets to improve democracy, the specific concern is that manipulators will choose to skew the markets at the time of some democratic event, such as an election or a vote in Congress. For instance, those who stand to gain from a president's stimulus package might bid up the conditional markets that show favorable economic indicators should the stimulus pass.

Even before evaluating how likely such manipulation is, it is again important to avoid the nirvana fallacy. In the absence of prediction markets, special interests and ideologues will try to make arguments that falsely claim public benefits for legislation or falsely deny them. Prediction markets at least balance such influence by providing incentives for actors whose motivation is accurately to predict consequences. In any event, manipulation is unlikely to be successful even in the short term when other traders are aware of its dangers, because they will be particularly sensitive to the opportunities for countering manipulation and making a profit.[48]

Formal modeling, experiments, and anecdotal evidence support the view that manipulators will face substantial difficulty. Two economists have offered a model showing that people who try to manipulate the markets offer higher returns for those who make accurate predictions.[49] In other words, because they are consciously trying to aim at the wrong price, the manipulators are sheep who attract wolves who are interested solely in accuracy. A set of experiments also suggested that manipulators who had incentives to distort the market from the payoff they would receive were unable to do so. Those without these incentives countered manipulators to produce accurate pricing.[50] Finally, past experience in electoral markets suggests that there are limits on the effectiveness of manipulation. Investors in political shares may have an incentive to bid up their candidates' price in the hopes of skewing the election by persuading voters to get on the bandwagon of the winning candidate. But spikes in electoral prediction markets that presumably represent either manipulation or exuberance on behalf of a candidate are short-lived, because counter traders bid the price down.[51]

It might be argued that such reasoning does not apply to legislation with large stakes, such as a capital gains tax cut, because manipulators can afford to lose a lot of money if their manipulations promote legislation that puts even more money in their pocket. But manipulation on a large scale means there are large scale profits to be made by bidding prices back to an accurate level. Hedge funds run by experts should be able to raise money to benefit from others' inaccuracy. Thus, attempts at manipulation seem more likely to create a more accurate market than a more inaccurate one. Accordingly, government policies that encourage more people to use prediction markets are likely the best way to discourage manipulation.[52] Moreover, companies running prediction markets have incentives to create rules that cut down on manipulation.[53] As I discuss below, government should support experiments in prediction markets that try to find designs to restrain manipulation.

The most serious issue for prediction markets is gauging the degree of confidence we should have in them. Given the accuracy of election markets, and the strong theoretical reasons to believe that markets are a good way of gathering dispersed information throughout society, there is a substantial case that prediction markets are likely to be better than other methods on making public policy predictions across the board.

Nevertheless, only a careful review of the past performance of such markets will help to confirm the circumstances in which we should have high confidence in their results and to discover any in which our confidence should be more measured. Empirical analysis, for instance, can evaluate how well information markets predict and how well their evaluations of future events track the distribution of events that actually happen. While such backward-looking analysis can never tell us exactly how markets will do in the future, it can provide useful rules of thumb to guide our use, giving us a sense of how confident we should be in their assessments in particular circumstances. Because accelerating technology will raise the stakes of collective decision making, it is important to gain a greater understanding of the uses and limitations of prediction markets now. Thus, the need for more experience with prediction markets on public policy and more experimentation with their design is itself a persuasive argument for a much friendlier government policy toward them.

## Legalizing and Subsidizing Prediction Markets

The most important policy recommendation for prediction markets is straightforward: remove the laws that impede their operation. Both regulatory and legislative changes are needed to free prediction markets from the constraints that prevent their development. Currently the Commodity Future Trading Commission (CFTC) applies regulations designed for commodity markets and derivatives. Lighter regulation is more suitable for these fledgling prediction markets.[54] The CFTC has already provided a harbor for the Iowa Electronic Futures Market, which has been running prediction markets on elections and Federal Reserve rates since the 1990s.[55] Unfortunately, that market has not expanded. Its relative inertia confirms the need for more competition among prediction markets in the profit sector as well as in the nonprofit sector. That competition can occur only if the CFTC provides more general permission with a regulatory structure that will permit such markets to flourish. Given that the CFTC recently has suggested that the Dodd-Frank law may prevent some kinds of markets in political results,[56] Congress may well have to take the lead in relaxing the agency's regulatory strictures.

Congress should also exempt for-profit prediction markets from federal gaming statutes. In late 2006 Congress passed the Unlawful Internet Gambling Enforcement Act (UIGEA), which banned financial institutions from transferring funds for the purposes of online gambling.[57] The law has generated a substantial chilling effect on the development of prediction markets. A number of gambling sites have pulled out of the U.S. market.[58] Creating another obstruction for prediction markets, two leading online payment operations—Neteller and Paypal—have refused to allow deposits to fund gambling accounts on the Internet.[59] The result has been to limit for-profit operations to offshore sites and to make it more difficult for Americans to bet on those sites.[60]

The laws are particularly infuriating because legal restrictions on prediction markets are not rooted in a genuine public policy against gambling. Federal law does not generally prohibit gambling; in fact, many states run large-scale gambling operations in the form of lotteries that provide no beneficial information for the public.[61] Prediction market restrictions provide an example of a so-called Baptist-bootlegger coalition, sustained by a minority of moralists who oppose anything that looks like gambling, and by some special interests, like horse-racing companies, who oppose legalized forms of competition.

But bad laws alone are not responsible for the paucity of public policy prediction markets. Even in nations like Ireland where such markets are legal, markets generally have been focused on entertainment issues such as predicting Oscar winners and horse races of both the equine and electoral kind. Conditional markets that benefit public policy are relatively few.

This development is also not surprising. Simple markets that concern celebrities and election results are likely to attract broader public interest.[62] A larger stream of visitors helps the online gambling sites earn additional revenue from advertising. Events like winning an Oscar or an election are also easy to reduce to a contract. Bettors would be deterred from markets on the long-term effects of public policy, because in the interim they would lose the time value of money.

Yet public policy markets, particularly conditional ones, help generate better information about policy effects of great significance to democracy in a manner that can catch citizens' attention. As a result, they provide a public good whose greater production society should support to a larger extent than the market supplies.[63] The federal government should thus undertake a program of subsidizing these markets. It should do so in an experimental way, seeking to find the market designs that are most accurate and the circumstances in which markets will be most useful.

For instance, the government could subsidize bettors to compensate them for some of the time value of the money they tie up in these markets until the accuracy of their prediction can be assessed. As a condition of receiving this benefit, participating information markets would have to agree to make their data available for empirical investigations like event studies. As a further measure of subsidization, the government could exempt payouts in conditional markets from some measure of taxation.

The government could also use subsidies strategically to encourage experiments in prediction market design. Short sales—the practice of selling shares one does not own—are often thought to make markets more accurate. The government could agree to subsidize one market on the condition that it permitted short sales and another on the condition that it did not and then evaluate the results. The federal government could also take advantage of federalism, providing grants to states to give tax breaks to encourage their own experiments with information markets on state public policy. Thus, yet another connection between empiricism and prediction markets is that government should encourage experiments in this important area.

The president of Columbia University recently argued in favor of subsidizing the mainstream media because of the advantages such subsidies might provide to public understanding and consequently to public policy.[64] But many media outlets mix opinion with fact. It is dangerous therefore to have the government choose which outlets to subsidize, potentially skewing public opinion. In contrast, prediction markets are based only on factual issues, creating substantially less risk that government subsidies will support some opinions over others. Of all the information technologies useful for social knowledge, prediction markets are the most in need of publicly supported experiments.

## Providing the Conditions for Markets through Government Transparency

Prediction markets require precise terms for the conditions on which bets are made as well as fixed measures to evaluate outcomes on which bets are paid off. This need for precision calls for another set of government information-eliciting rules—requirements that legal proposals be made publicly available before further action is taken on them. Thus, before a legislative committee votes on proposed legislation, before either chamber votes on a bill on the floor, and before the president signs a bill, the exact language at issue and all relevant amendments should be posted for everyone to see. Prediction markets can then be made on more exactly specified conditions. These timetables will also facilitate other mechanisms for promoting social knowledge. Experts can analyze the provisions and inform the nation about the likely consequences. Blogs can publicize hidden special interest provisions in a bill and refine the arguments for and against that bill before it is passed or signed.

Such a requirement would provide a legislative analogue to requirements that force government agencies to make their regulatory proposals available for a substantial period before they can be promulgated.[65] The time period for such analysis would not have to be lengthy. Prediction markets can be established quickly, experts can opine expeditiously, and blogs will open debate immediately. Useful requirements could be as short as a week or ten days.

Such rules should have exceptions for national security emergencies. Moreover, these requirements should not be confused with efforts to force Congress to open all of its deliberations to the public. Open meeting rules reflect the romantic notion that broader participation at every stage in the democratic process provides a better legislative product. The focus instead should be on better evaluation of well-defined proposals by the most knowledgeable people. Nor do such rules require that every institution Congress creates operate in public. Some institutions, such as a commission that closes military bases, may work better if they operate in secret. But the virtues of such operational secrecy need to be evaluated publicly at key points in the legislative process.

As is the case with some other reforms recommended here, the body politic is already sensing, however inchoately, that such rules would be beneficial. President Obama promised during his campaign to wait for five days before signing any nonemergency bill.[66] Although he has not fully honored that commitment,[67] his promise reflects the public resonance of greater legislative transparency. The current House of Representatives has required that a bill be posted three calendar days before a vote.[68]

Prediction markets require measures of public policy results, such as estimates of economic growth, that are free from manipulation, because the measures will determine payouts. As these indicators will also often be compiled by the government, their use becomes yet another strong argument for transparency and stability. Government should be free to create new measures of such matters as economic growth, but it should continue to calculate old measures so as to permit information markets to anchor their predictions.

## Prediction Markets as a Response to Social Complexity

Technological acceleration has made for a more complex society. As a result, collective decision making faces more complicated policy-making options than ever before. But technological acceleration has also provided us with mechanisms to address this greater complexity. By highlighting the actual effects of specific policies, prediction markets can help separate out the analysis of their actual consequences from their proclaimed goals. As a result, they can test whether politicians' programs are likely to fulfill their stated objectives. Like government support for empiricism, support for prediction markets can help foster a politics based less in pretense and more in facts. As with the empirical analysis of social policy, however, the government must adopt a variety of information-eliciting rules to make this new mechanism more effective.

The longer-term effects of prediction markets are, perhaps paradoxically, hard to predict. It is difficult to foretell all the synergies that prediction markets may create with other forms of analysis that are driven by more powerful computation. Prediction markets are a classic emergent platform of innovation—this one in policy analysis—and the connections that such broad platforms make with other enterprises in the same field depend on the imagination of their users. But the prospects of improving policy analysis through such markets are good enough that they are worthy of encouragement, not harassment, by the government.

Prediction markets will have greater effect on the polity if their results are more widely discussed. At least initially, these novel markets are likely to be discussed by blogs and other specialized media before gaining a foothold in the conversation conducted by the more mainstream media. This mainstreaming is already beginning: sophisticated journalists at publications as diverse as the *New York Times* and *National Review* referred to electoral prediction markets during the 2011–2012 Republican primary season. It is thus to the new information technology of dispersed media that we now turn.

# CHAPTER FIVE

# Distributing Information through Dispersed Media and Campaigns

EMPIRICAL INVESTIGATION and prediction markets can generate substantial data about the likely effects of policy. But by themselves they cannot broadly distribute the data to the general public. Moreover, by themselves they cannot create debate about explanations of the relevant facts. Other information technology must be responsible for distributing this information and for creating platforms for debate. Fortunately, new information technology has also created a more dispersed media that has the capacity to bring a sharpened deliberation to the data and set the stage for a more intensive discussion of policy results in political campaigns.

## Dispersed and Innovative Media

The most beneficial effect of the Internet on democracy is its capacity to produce a better evaluation of policy consequences, thereby advancing a politics of learning. Because of the greater space and interconnections that the Internet makes available, Web-based media like blogs can be dispersed and specialized and yet also interconnected and connected with the wider world. As a result of this more decentralized and competitive media, the Internet generates both more innovative policy ideas and better explanations of policy than were available when mainstream media, dominated the flow of political discussion. But the more mainstream media, like newspapers and television networks, remain an important part of the media mix, distilling the best of the Internet and bringing it to a wider public.

Specialized blogs can address policy issues with a level of sophistication and depth that the mainstream media never could achieve. In my own field of law, the Internet hosts scores of widely read blogs, often with a quite particular focus, from tax law to contracts, from empirical studies to law and economics.[1] Websites like Marginal Revolution, Calculated

Risk, Grasping Reality with the Invisible Hand, and the eponymous blog of Harvard professor Greg Mankiw[2] offer distinctive perspectives on economic policy.[3]

Such specialized media improve our factual knowledge of the policy world in two respects. First, specialized media deepen substantive knowledge by providing a framework for the presentation of information in particular policy areas.[4] Experts participating in specialized media that are closely followed by their peers gain greater incentives to explain policy effects carefully and with nuance. It is not that they will all agree, but they will try to avoid obvious mistakes and respond to counterarguments. In this world of intense scrutiny, scholars will suffer blows to their reputation if they fail to engage in careful self-monitoring.[5]

A signal example of expert monitoring occurred recently in one of my own specialties, constitutional interpretation. A leading advocate of interpreting the constitution according to its original meaning engaged in a dispute with a historian about posts he put up discussing the original meaning of some clauses of the Constitution.[6] The back and forth became somewhat heated. But by the end, a variety of law professors, most of whom were not advocates of originalism themselves, intervened to complain about the historian's tone and failure to grapple with the argument on the merits.[7] Thus, the expert blogosphere polices itself.

This disciplined presentation of specialized information has the potential to better match an interest in truth seeking with an individual interest in personal advancement, because experts leave a continuous record of their policy claims and explanations both for themselves and for the interested public. As Philip Tetlock, an expert on the nature and effect of expertise, observed, "The more people know about pundits' track records, the stronger the pundits' incentives to compete by improving the epistemic (truth) value of their products."[8]

Blogs also improve the methodology of substantive analysis as well as the debate on the substance of policy.[9] As social science advances, it depends on more complex mechanisms for gathering and analyzing data than simple observations. This process of resorting to more complex and theory-laden methods for substantive analysis is analogous to that used in the natural sciences. Astronomy no longer depends on the evidence of the senses, but on tools like telescopes and computers whose operation and output rest on scientific theories. Nevertheless, these tools give us a far better understanding of the cosmos than simple observation can provide.[10] Indeed, direct observations of the senses can be wholly misleading. Stars do not actually twinkle, although that is how they appear to observers on earth.[11] Similarly, simply looking at homicide rates in Massachusetts and Idaho would not tell us much about the effects of gun control unless we used more complex methods of observation and analysis.

The more complex methods of social science are consonant with broader human progress. But this very complexity may lead to a greater danger of methodological mistakes and even ideological manipulation. As such, these methods require theorizing and criticism to perfect them, just as theorizing and criticism are needed to improve substantive explanations of the social world. Dispersed media can help with this enterprise as well.

Second, dispersed media help provide better information to the general public. As blogger and law professor Glenn Reynolds has noted, such specialized media feed into the more mainstream media, providing the kernels for stories that reach a wider public than those who follow the give-and-take of blogs.[12] A modern reporter's beat is not simply pounding the pavement, but wading through the Web as well.[13] For instance, the blog Economix on the website of the *New York Times* regularly links to professional economic analysis, often offering facts in tension with factual claims supporting that newspaper's editorials.

The Internet has become a vast funnel of information that allows specialized but politically salient information to be diffused into the wider world in a form in which it can be readily understood.[14] Hypertext makes these connections visible: the reporter links to the academic article on which he relies, just as the academic may link to the data set on which she relies. The funnel has a two-way flow as well; such dispersed media also check more mainstream media, creating another kind of constraint that makes the information flow more accurate.[15] Thus, the dispersed media create a better discovery function for deciding which issues should be on the polity's agenda. No longer do we rely on the relatively unguided judgment of journalists. Instead, there is a more direct pipeline from experts. This greater degree of sifting and expert influence is important, because the electorate does not itself patrol for issues but relies on the media to do so.[16]

More generally, the funnel created by the interplay between specialized and more general media helps perform two crucial functions in structuring knowledge for social use. First, it helps with a concern that goes back to Aristotle: how does one elicit the technical knowledge that experts have and make it available as social knowledge that can be used to facilitate social decision making? The refraction of more specialized media through mainstream media makes it easier for the public to access the available technical knowledge developed by social scientists and other experts. Second, making this knowledge available to a broader public creates greater common knowledge. Common knowledge plays an important role in social decision making, because it creates positive spillovers. Knowledge in many minds is greater than the sum of the knowledge in individual minds, because it permits recombination for even more ideas.[17]

The ongoing switch from television to more dispersed media also improves the nature of common knowledge. Television emphasizes the personal, encouraging people to make more than warranted extrapolations from the personal characteristics of a candidate to the improvements he or she will make to public policy.[18] But the new, dispersed media encourage a more policy-oriented evaluation of the candidates' positions on the issues, because they are less personal and more focused on issues.

Some critics have expressed concern that the Internet and other new media actually represent a danger to democracy because they will lead to more overall polarization, blocking agreement and progress on social issues.[19] Legal scholar Cass Sunstein is the most prominent of these theorists. His warning that the new media may undermine democratic deliberation combines a theory about the Internet with a theory about human psychology. He argues that the Internet filters out dissonant information, because people tend to live in an information bubble where all the information reinforces their worldviews. So conservatives visit only right-leaning sites and liberals visit only left-leaning sites. He relies on psychological experiments that suggest that individuals embrace more extreme views when they associate with other extremists and are exposed to more arguments on one side than the other.

Both steps of Sunstein's argument are open to challenge. First, as a matter of theory, perfect filtering out of other views is not possible, because perfect filters have not been devised.[20] As a matter of fact, there is little evidence that even the better filtering afforded by the Internet has given people a more monochromatic view of the world. One recent study has suggested that ideological segregation is not severe.[21] Conservatives feed on a diet of online information that is the ideological equivalent of reading *USA Today*, a relatively middle-of-the-road publication, and liberals thrive on the equivalent of listening to CNN, a relatively middle-of-the-road network.[22] An individual's online community is actually less ideologically segregated than his or her neighborhood or network of friends, suggesting that the Internet may expose people to more diverse views than they would otherwise experience in the offline world.[23]

Nor does the Internet provide the same context as the experiments on polarization on which Sunstein relies. These experiments offer participants an intense experience over at most a few days. People tend to graze intermittently on the Internet and intersperse their online experience with their daily life. This difference weakens any extrapolation from the experiments to conclusions about the Internet's effect on ideological extremes, because people's understanding of the world reflects a mix of views perused on the Internet, their interactions with others in daily life, and their reaction to basic current events. In any event, it turns out that those who are online regularly are more tolerant of ideas they do not like than those who are not.[24]

Sunstein's view also ignores the potential of new information technology to put greater emphasis on facts than on ideological opinions. Empirical studies of past policies focus on whether those policies actually work. Prediction markets can help assess what the policies will actually achieve. The Internet is not being created in isolation, but as part of a technological revolution in information that reorients the world's landscape, impeding the likely success of a worldview that does not confront facts that may be in tension with it.

The very structure of the Internet tends to encourage transparency in the factual assumptions of arguments, and this transparency invites dialogue and refutation. Hyperlinks provide an easy way to ground one's own argument in facts and point out the errors of others. More generally, when one side makes factual claims that are crucial to its argument, the other side has incentives to show that those claims are not true. As a result, the World Wide Web creates a more universal and ubiquitous forum in which all can contest policy claims.

Two recent examples can make this abstract point more concrete. Paul Krugman, a Nobel Prize–winning economist, argued that the United States should not fear European levels of taxation, because the growth rates in Europe per capita were only slightly lower than those in the United States.[25] Many other economists responded to his blog post, providing additional facts that rendered this analysis problematic. One noted that the United States was already substantially wealthier than Europe, and thus the continuing faster growth in the United States was even more telling, because it is easier to have higher growth rates at lower levels of wealth.[26] Another noted that the United States has a much higher per capita income than almost all European nations, thus making clear that even small differences in yearly growth rates could result in a large difference over time.[27] A third observed that a more telling comparison turned on how different immigrant groups fared in the United States compared to their fellow countrymen left behind in Europe—a comparison very favorable to the United States.[28]

Robert Barro, a more right-leaning Harvard economics professor, who is frequently mentioned as a candidate for the Nobel Prize, recently published an op-ed in the *Wall Street Journal* arguing that extending the unemployment benefits during the 2008–2009 recession had substantially increased the unemployment rate.[29] Undeterred by his prestige or authority, bloggers raised substantial doubts about the factual underpinning of his argument. A key premise of Barro's claim was that it is possible to extrapolate what the employment numbers today should be from the employment rate during previous recessions. But commentators observed that the recent recession was very different from past recessions because of huge losses stemming from a housing bubble.[30] These losses in turn depressed new housing starts, which had accurately predicted the unemployment rate. Another commentator provided evidence that Barro had gotten the relationship between unemployment insurance and unemployment backward.[31] It was high unemployment that caused the duration of unemployment insurance to be lengthened rather than the other way around.

Factual confrontations such as these generate substantial advantages for democratic updating. They help winnow out the more plausible factual contentions from the less plausible. Just as scientific and technological innovation occurs from exchange and disagreement, so does policy innovation. Through exchange some ideas are discarded and others are recombined to yield new approaches.

The mechanisms for updating are less rapid and unerring in policy innovation than in technological innovation. In technological innovation, inventors and entrepreneurs ultimately make more money by discarding less productive ideas. Nevertheless, as I will describe in more detail in chapter 8, even in the democratic process experts have incentives to update their views on the basis of new information because of the effects on their reputation, and politicians have incentives because of the effects on the outcomes of elections. Since these incentives are much more modest and less direct than monetary ones, updating will take longer and be more imperfect, but it still can be effective over the long term.

Thus, while there is no doubt that a number of very partisan media outlets exist on cable TV today, partisan media are nothing new in American history. At the beginning of the republic, political parties even controlled newspapers.[32] While many newspapers became more objective in the twentieth century, partisan outlets continued even then. What is new is the capacity of dispersed and specialized media to inject a continuous stream of more disciplined expertise into political discourse—a stream that other new technologies, like prediction markets, and old technologies, like the more mainstream media, can turn into a powerful flood.

Some people wax nostalgic for a time when a relatively few established sources—the major networks and major newspapers—set the agenda for social policy through their decisions about what to report.[33] But just as more vigorous market competition improves consumer welfare by creating better products, more vigorous competition in ideas should improve policy, at least when the medium of argument itself creates a trail to the factual grounding of arguments.

The information advantages of dispersed media over more concentrated media are similar to those of democracy over oligarchy. Other forms of government appear much more stable than democracy, because they create less surface conflict. But the absence of conflict makes it more difficult to find policies that will work. Media oligopoly, like that of the era when three television networks dominated the news, is not dictatorship, but the relative absence of factual confrontation projects a pretense of social knowledge that is not well rooted in actual knowledge.

The acceleration of technology in particular requires the greater contentiousness generated by dispersed media, because such technological change may well call for faster changes in the status quo. It is important to have a media dynamism that brings diverse ideas into a confrontation of ideas and refines them faster than was possible in the more static media of the past.

## A Culture of Learning

The information technologies discussed so far—empiricism, prediction markets, and dispersed media—have the potential to create a new democratic culture of learning, focusing on the consequences of policies. They create synergies, providing improvements that are more than the sum of their parts. Empiricism gathers information about the past that helps prediction markets test the future. Prediction markets discipline the bias of empirical experts, providing a forum for the essential mixing of expert knowledge and the more dispersed knowledge of citizens in general. Dispersed media bring to the fore issues that should be on the political agenda for testing and prediction. They can operate more rapidly than empiricism and more flexibly than prediction markets, showing that another advantage of these technologies is their complementary nature. Further, in regard to social events, dispersed media offer hypotheses that can be tested. If facts without theory are blind and theory without facts is inert, the mix of new information technologies and the recombination of ideas they create help generate a needed synthesis.

Long ago political scientist Charles Lindblom saw a kind of informal process of policy falsification as the essence of advanced democratic practice; democracy provides "a piecemeal process of limited comparisons, a sequence of trials and errors followed by revised trials."[34] But in today's world of technological acceleration, democracy needs to reduce its errors because of the larger costs of mistakes with possible catastrophic consequences as well as possible huge benefits forgone. The new information technologies speed the trial-and-error process along.

So long as government adopts better information-eliciting rules, these developments in information technology can improve political culture more generally by encouraging greater attention to the consequences of policy. Even with the right rules, the changes wrought by these new technologies to democracy will be gradual in the short run, but over time they can make a large difference. Indeed, as more fact-based sources of information become more prevalent in political life, we might expect some people to embrace an empirical political culture as part of their identity, focusing on the latest predictions from information markets and empirical studies and leaving behind more partisan or ideological worldviews as central to their sense of self.

Of course, some people will continue to hold to ideologically based beliefs, even while proclaiming that they are open to evidence. But the dynamic of better information puts such political actors in a box. Better evidence undermining their factual claims forces them to rely on naked ideology, a stance that is rarely persuasive to the center and is thus ineffective in the long term in a democracy.

Some might regard as impoverished the kind of democratic decision making that empirical studies, prediction markets, and fact-based dispersed media will promote. This kind of democracy focuses on instrumental questions, such as what will create economic growth, improve health care outcomes, or decrease certain pollutants. Consequentialist democracy is of far less help in deciding more cosmic social questions, such as whether economic growth is a good thing, or moral questions, such as whether abortion should be legal. But there is little evidence that the public at large is preoccupied with the cosmic questions, and although many people are intensely interested in social and moral questions like abortion, democratic discussion does not seem to promote either consensus or resolution.

It might be objected that by sidestepping the enduring questions of meaning and human identity, however intractable, empirically based democracy is a political system unworthy of our deepest concerns and longings. But even if a consequentialist democracy does not focus on these issues collectively, citizens are free to address them privately and indeed try to change the consensus of society by making normative arguments. And a democracy that scrutinizes consequences achieves a goal that seems fully worthy. It improves human life by helping society make the policy choices that will most efficiently and least counterproductively realize its consensus values.

## Assuring the Distribution of Information

Dispersed media, particularly when combined with empirical inquiry and prediction markets, have the capacity to create a politics more focused on the consequences of public policy. Just as the government in the nineteenth century helped distribute policy and political information through the post office, so today it should be careful to facilitate distribution of such information through contemporary technologies.

First, our laws should give as much protection to the new, dispersed media as to the old media. Second, the government should also encourage universal access to the Internet—the portal to much of the dispersed media. Finally, the government should deregulate and subsidize the provision of information in political campaigns, because campaigns remain the most effective route for public policy information to reach the mass of citizens who do not follow specialized media or even the news more generally. It is expensive to get the attention of inattentive citizens. Government thus needs to relax its strictures on individual contributions to political campaigns and subsidize the contributions of people of modest means to encourage their greater participation. Otherwise, the fruits of the information revolution will not be disseminated as broadly as they can be.

## Avoiding Regulations that Discriminate Against the New Media

The value of dispersed media in a democracy underscores the importance of preventing regulations that discriminate against such media. Laws that shield reporters of traditional media from libel judgments and the obligation to produce notes to grand juries should also apply to protect bloggers. For instance, it was wrong for a court to deny a blogger the privileges of an Oregon law that requires plaintiffs to first request a retraction from journalists before suing for libel.[35] Any discrimination between old and new media may raise First Amendment concerns, because that amendment should apply equally to everyone, regardless of size or kind of media.[36] But besides its constitutional infirmity, such discrimination is a policy blunder. Even if blogs and other new media are individually small, cumulatively they play a substantial role in educating the public. To treat them as less worthy of protection reflects a failure to grasp that such rivulets are now becoming the stream that nourishes public deliberation.[37]

Blog postings also should be no more heavily regulated by campaign finance rules than editorials.[38] Any discrimination in favor of the old media would decrease the diversity of information, because the new media are likely to bring specialized information to the attention of citizens either directly or through its republication in more traditional media. It is precisely near elections that those with empirical data and expertise are most needed to critique the policies and platforms of candidates.

## Encouraging Universal Access

Just as the federal government in the nineteenth century tried to provide access to information by subsidizing the U.S. Post Office and periodicals that were sent through it, today it should subsidize access to broadband Internet. It has been argued that such subsidies help economic growth by facilitating the recombination of entrepreneurial ideas.[39] But it is also important to facilitate the recombination and distribution of ideas about policy. Besides making access affordable, the Obama administration should continue its laudable efforts to try to improve basic computer literacy among the population by sending teams to libraries and stores in disadvantaged and rural neighborhoods.

## Getting Information to Voters

New information technology is generating more information about issues of public importance. Empirical inquiry can evaluate past policy, and prediction markets can assess future policy. Dispersed media constantly scrutinize and debate the main currents and finer points of our social agenda.

Some people will follow this new cornucopia of information out of personal interest or civic duty. Most, however, will pay only intermittent attention to politics or policy. Indeed, accelerating technology exacerbates the problem of public inattention, because the same technology that creates more information about public policy distracts citizens from the business of public life with the greater variety and richness of entertainment. Thousands of people were willing to attend the Lincoln and Douglas debates. The crowds should not surprise us, because the debates were the best show in town.[40] The pleasurable pursuits of today are vastly more numerous than those existing in 1858. Politicians must compete with hundreds of cable television channels, featuring people far more attractive than most of them and engaging in activities most citizens find far more amusing. Thus, even if the physical cost of distributing information about politics is declining with new technology, the cost of catching most citizens' attention may be increasing.[41]

An age of accelerating distraction as well as accelerating information calls for better rules to assure the maximum dissemination of political information. Political campaigns remain the best route for providing information to citizens who otherwise pay little attention. First, voters do pay more attention to candidates and public policy when the campaign season rolls around.[42] Second, politicians have powerful incentives to get information to voters during their campaigns.[43] Of course, they also have incentives to distribute information that is slanted in their favor, but their opponents have incentives to dispute any false or misleading information. It is an imperfect system, but democracy has not devised a better way of making the mass of citizens better informed.

The level of a candidate's campaign spending has a direct impact on the public's level of knowledge about that candidate. Researchers have found that the more a candidate spends, particularly in the form of advertising expenditures, the more informed the citizenry becomes.[44] In response to increased campaign spending, citizens are better able to place candidates on an ideological scale and to know what their positions and votes are. Indeed, one scholar has found evidence that campaign spending in the form of advertising has a favorable impact on voter knowledge that rivals that of television news.[45] Much advertising is negative, but such advertising brings to the public's attention the personal and policy downsides of electing a particular candidate.

Other more indirect evidence also confirms that people become more informed as campaign spending increases. Challengers increase their vote share the more they advertise.[46] Challengers' advertisements make more of a difference to their vote shares than does the advertising of incumbents, presumably because incumbents are better known than challengers.[47] More advertising on judicial elections prevents runoff—the practice by which voters do not vote on candidates farther down the ballot.[48] If such advertising did not increase knowledge, voters probably would not bother voting for judicial candidates in whom they had little confidence.

Thus, political advertising does help make people more informed about the election, both the identity of candidates and their policy positions. We do not have an empirical study showing that elections make citizens more directly familiar with policy consequences. Nevertheless, such advertising is essential to help voters connect whatever familiarity with policy consequences they have to the candidates who are responsible for implementing policy. Knowledge of policy consequences is inert in elections without knowledge of candidates' stances on policy.

Of course, much political campaign advertising, like much advertising in general, makes less than rational appeals. But much political advertising already provides some information about consequences. Many commercials offer facts about the votes of representatives on bills of interest and relate them to economic conditions or other consequences. For example, President Ronald Reagan's "Morning in America" campaign—one of the most famous and effective advertisements in modern American political history—focused on facts about the improving conditions in the U.S. economy with the clear implication that Reagan's still controversial policies were responsible for them.[49] In contemporary legislative races we often see a focus on policy consequences. A Democratic candidate's advertisement in 2010 noted that proposals to privatize portions of Social Security previously supported by his opponent would have exposed people to substantial losses in the stock market in the 2008 recession.[50] A Republican focused on the cost to taxpayers of President Obama's stimulus package and its failure to achieve the president's proclaimed goals.[51]

The dynamism of new information technology is likely to increase the capacity of campaigns to focus effectively on the consequences of policy.

As the funnel of information generated by dispersed media brings more salient facts to bear on public policy, we can expect more such facts to be used. Advertisements for personal products frequently focus on showing that products work for consumers as a way of creating differentiation from other similar products.[52] There is no reason to expect that political advertisements will not incorporate more information about which policies work as it becomes available, assuming that a candidate has money to make these claims and contest the claims of his or her opponent. For instance, if a mayor executed a plan that raised educational achievement according to respected empirical studies, he will trumpet it, just as a toothpaste company touts the proportion of dentists who recommend its product. Similarly, if a president has a tax program that prediction markets suggest will increase economic growth, he is likely to make that prediction known. Thus, it is important to understand that ultimately political campaigns will be strongly affected by the ecology of the society's entire system of information production. Because of the potential of empiricism, prediction markets, and dispersed media, with the correct policies that ecology can become more friendly overall to consequentialist argument.

Some political advertisements, like commercials for consumer products, are migrating from television to the Web. We can expect evolution in the form as well as substance of advertising to continue as information technology evolves. Perhaps we will have campaign advertisements in the form of interactive videogames to draw people in to the debate. Whatever form they take, of course, political commercials will never deliver information with the precision of social scientists. But the information that they convey, not the information that they fail to convey, is the most important fact about them. The alternative to political advertising is not a policy seminar, but a beer commercial.

To be sure, the effectiveness of campaign advertising to inform the public varies depending on other factors. Better educated and more knowledgeable voters seem to benefit more from advertising than others.[53] But, as will be discussed in chapter 8, democracy can update when only some citizens update on the facts. If advertising improves the knowledge of a substantial number of people, it can help improve collective decision making.

Not surprisingly, people pay more attention to an election the more closely contested it is.[54] Thus, the uptake of political information will generally improve if more candidates run in districts where the principal parties are relatively evenly matched. This observation provides further support for reforms that would prohibit partisan gerrymandering— legislative districting that is often designed to assure the election of members of particular parties (reforms that will be discussed in chapter 9).

## Providing More Money for Campaigns

Despite the likely effectiveness of campaign advertisements, campaign expenditures are relatively small, particularly given the large expenditures that candidates will control upon election.[55] The amount spent on advertising for political campaigns is orders of magnitude less than advertising for similar expenditures on private products.[56] To put this discrepancy in some perspective, in 2004 private sector spending through direct mail alone totaled $52.19 billion,[57] while in 2010 House Democrats and Republicans raised a total of less than $1.1 billion.[58] Thus, the constant complaint that too much money is spent on elections is mistaken.[59] In fact, this belief is more misplaced than ever in an age of accelerating technology when the stakes of decisions may become higher. Spending on politics offers more information about policy direction and results and is essential to connecting policy stances to candidates.

The two best ways to increase the amount of money available for political spending on advertising are to relax the current limits on the contributions of individuals to candidates and parties and to provide tax credits for citizens of more modest means for similar donations. These policies would help provide more money for political advertising without substantially empowering special interests.

The most obvious way to provide more money for advertising is to permit larger individual contributions to candidates and political parties. Currently at the federal level the limitations on contributions for each citizen are $2,500 per candidate and $30,800 to a political party.[60] The only rationale accepted by the Supreme Court for limiting individual contributions is fear of corruption or perception of corruption. But the notion that in an economy the size of the United States a citizen can corrupt a member of Congress for contributions of even four or eight times the current limit is implausible.[61] Given the wide variety of interests represented by a party, the ceiling for contributions to political parties should be substantially higher. Moreover, given modern information technology, disclosure seems a more powerful limitation to corruption than ever before. Dispersed media can make clear the sources of contributions and connect the dots between contributions to the policy positions that directly or indirectly benefit individuals who make contributions.

If Congress fails to raise contribution limits, the Supreme Court should strike down the current low limits. The Supreme Court has long played a leading role in its interpretation of the First Amendment, the most important constitutional provision on the distribution of information. This action would be consistent with a jurisprudence encouraging social discovery, because other information-eliciting rules will become more effective if information gets to the voters.

Also consistent with an information promoting jurisprudence, the Court has upheld disclosure of campaign contributions.[62] Congress should require that disclosure be immediate, transparent, and posted on the Web. One of the unfortunate consequences of the crazy quilt of campaign contribution laws and regulations has been loopholes that permit candidates and independent groups to hide the identities of their contributors.[63] Congress should make clear that shell corporations and other sham methods of avoiding disclosure are illegal. It should also increase the penalties for disclosure failures. In this way there will be both money for candidates to make their case to the public and information about the sources of that money. The Federal Communication Commission's recent direction to broadcasters to make available online data about political advertising is a step in the right direction.[64]

The Court should not deal more leniently with state limitations on contributions to state campaigns. In general, state experimentation with different programs is to be encouraged so that policies can be compared and policy progress can be made. But knowing the results of policies is a precondition of that progress. As a result, protecting information flows in the course of political campaigns is one area where the Supreme Court should not create as much space for state experimentation.

There are other rationales for limiting individual donations to campaigns, but they have not been endorsed by the Supreme Court and are even less persuasive than a concern about corruption or its appearance. It is said that more money will allow more special interest influence, thus distorting collective decision making;[65] special interests can indeed be a source of bias. But raising the level of giving by individuals in an election is unlikely to provide more resources to special interests. Special interests gain power when they can overcome collective action problems and direct large sums for common purposes by binding their members to donate. Trade associations and unions are prime examples. But even if individuals could contribute ten times the current limits, they cannot as easily wield concentrated influence on behalf of particular projects or regulations, because they have no legal mechanism to bind others to direct their contributions for a common purpose.

Indeed, raising the maximum level of individual contributions would empower individual citizens, not organizations. Thus, even if one accepts the controversial argument that contributions affect legislative voting behavior, this reform would make classic special interests like trade associations and unions relatively *less* powerful, because the contributions they make and the independent funds they expend in an election would become a smaller portion of the overall level of political support.[66] Politicians would not depend so much on fund-raising by lobbyists and insiders, because they could raise large sums from people across the nation.

Even the power of special interest funding is self-limiting.[67] An empirical study has suggested that if the contributions come from special interests, the message so funded is less persuasive, presumably because using special interest money makes the candidate less credible.[68] New information technology is likely to increase this effect by making it easier to show which candidates are relying on a narrow base of support. Candidates who receive most donations from interests seeking particular projects or favors will receive less bang for their advertising buck than candidates who raise their money from individuals promoting their perspective of the public good.

One concern is that the recent Supreme Court decision in *Citizens United v. FEC* would require that for-profit corporations be permitted to contribute the same amount as individuals.[69] It is true that in *Citizens United* the Court held that Congress could not ban independent expenditures by corporations. (Independent expenditures support electioneering advertisements that are not coordinated with a candidate.) But since 1907, the law has banned direct contributions by corporations to candidates, and this ban was upheld long ago.[70] Making a distinction between individual and corporate contributions continues to be as reasonable today as it was when the original law was passed. For-profit corporations would be violating their fiduciary duty to shareholders if they were not trying to get something tangible in return for their contributions,[71] whereas individuals can and do have interests in the well-being of the nation for which they expect no material recompense. Thus, while there may always be an appearance of corruption in the contributions of for-profit corporations to politicians, a contribution by a citizen does not raise the same level of concern.

It has also become clear that Citizens United has not actually resulted in a large increase of independent expenditures by for-profit companies, no doubt because most corporations do not want to annoy potential customers or employees.[72] The most important effect of Citizens United has been to allow citizens to band together in non-profit corporations, so-called "Super PACs," that make independent expenditures on their joint behalf. But individual citizens acting together are not presumptively special interests. They have no legal mechanisms to force others to pool resources with them. Citizens intervene in politics not necessarily to seek private gain but to persuade others to follow their vision for the nation, disseminating information about politics and policy in the process.

It is also said that large individual contributions or independent expenditures distort the equal influence that each person should wield in a democracy.[73] The equality concern has the most bite if the only important function of democracy is to elicit preferences, because by their very nature the additional contributions by the wealthy may be thought to make politicians focus more on their preferences. Permitting rich people to donate more money, however, is less likely to substantially distort democracy's function for evaluating consequences, because consequences are facts about the world that are independent of preferences. Money is needed to uncover these facts and bring them to the attention of the public.

One might have reason to be concerned if the wealthy had a single set of beliefs about likely consequences and contributed only to candidates who supported policies that accorded with those beliefs. But in our democracy the wealthy have a diversity of political viewpoints. Both of our principal parties have very substantial support among the wealthy. In the 2008 election, among those with incomes over $250,000 per year, more people voted in favor of Obama than McCain,[74] even though Obama ran on a platform of raising their taxes.

Indeed, reducing money in politics may make it more difficult to get an unbiased view of policy consequences. In the absence of substantial campaign advertising, groups other than the wealthy who put out information on policy, such as journalists, academics, and Hollywood celebrities, will have more influence. These groups lean much more heavily to one side of the ideological spectrum than do the wealthy.[75] Campaign restrictions also make the press unduly powerful. Britain has severe campaign finance restrictions, and it is thus not surprising that both the Labour and Conservative parties have spent a lot of time currying favor with press barons.[76] More generally, because the wealthy gain their income from a vast variety of enterprises whose interests often conflict, it is unlikely that they will have as monochromatic views as other groups poised to influence politics, such as the press, entertainers, or educators.[77]

Providing a tax credit of a few hundred dollars for political contributions to candidates can temper equality concern. This credit should phase out at higher income levels, because higher-income people can afford to contribute on their own. The credit would allow people of modest means to donate to their favorite candidates or parties. Such a tax credit would also inject more money into political campaigns and thus help get relevant policy information out to voters.

From 1972 to 1986 a federal political contributions tax law existed under which American taxpayers could claim tax credits for contributions to federal, state, or local candidates.[78] This program provided two options: taxpayers could claim a 50 percent tax credit or a 100 percent tax deduction for contributions to candidates, with an annual maximum benefit per taxpayer.[79] Providing a 100 percent tax credit limited to people of modest incomes should garner far more participation than this previous program, since those receiving the credit could contribute without ultimately being out of pocket.

Tax credits for candidate contributions provide a better way of injecting more money into political campaigns than direct public financing in which those candidates receive public money based on some set of criteria, such as matching private contributions. First, the tax credit is superior on equality grounds. It provides the appearance and reality of more participation by lower-income individuals in deciding who gets support. Tax credits also do not require the government to determine who satisfies the criteria of public financing. Nor do they require as large a bureaucracy to monitor the disbursement of public funds to assure they are being spent properly on campaigns. As with private charitable contributions, individuals act as the front-line monitors, because they are likely to focus on whether the money they contribute is devoted for the purpose for which it is sought. Finally, tax credits are simply more politically salable, because empowering individuals with tax credits is more popular than creating a taxpayer-supported fund from which the state doles out money to politicians.[80]

It is a mistake to object to such tax credits because of a belief that they add to government spending. Just as the government should subsidize prediction markets because they provide valuable information that constitutes a public good that would otherwise be underproduced, so it should subsidize the delivery of information at the time of election, because that valuable good is also otherwise underproduced. The new information provided by technology will make more of an electoral difference the more widely it is disseminated and the more it is connected to the choice of candidates at election time. The result of such spending in the long run should be more efficient and less counterproductive government programs, as the electorate is better informed of the connection between candidates and the consequences of policy decisions they support.

# CHAPTER SIX

# Accelerating AI

MANY DIFFERENT KINDS OF technologies, from nanotechnology to bio-technology, promise to dramatically change human life. But of all these potentially revolutionizing technologies, the most important for social governance is artificial intelligence (AI), because AI is an information technology. As a result, the development of machine intelligence can directly improve governance, because progress in AI can help in assessing policy consequences. More substantial machine intelligence can process data, generate hypotheses about the effects of past policy, and simulate the world to predict the effects of future policy. Thus, it is more important to formulate a correct policy toward AI than toward any other rapidly advancing technology, because that policy will help advance beneficial policies in all other areas.

The holy grail of AI is so-called strong AI, defined as a general purpose intelligence that approximates that of humans. Strong AI has the capacity to improve itself, leading rapidly to machines of greater-than-human intelligence. More concretely, it raises dramatic possibilities of very substantial benefits and dangers, from the prospect of a machine-enabled utopia to that of a machine-ruled despotism.

Fortunately, the correct policy for AI—substantial government support for Friendly AI—both promotes AI as an instrument of collective decision making and helps prevent the risk of machine takeover. Friendly AI can be defined broadly as the category of AI that will not ultimately prove dangerous to humans.[1] The benefits of supporting Friendly AI are two-fold. First, the creation of Friendly AI is the best and probably only way of forestalling unfriendly AI, because Friendly AI can prove a crucial ally to humanity in its efforts to prevent the rise of dangerous machines. Second, government support is justified on the wholly independent grounds of improving social decision making. Even if strong AI is not realized for decades, progress in AI can aid in the gathering and analysis of data for evaluating the consequences of social policy, including policy toward other transformative technologies.

## The Possible Coming of Strong AI

The idea of artificial intelligence powerful enough to intervene in human affairs has been the stuff of science fiction from HAL in *2001: A Space Odyssey* to the robots in *Wall-E*.[2] The notion of computers that rival and indeed surpass human intelligence might at first seem to be speculative fantasy rather than a topic that should become a salient item on the agenda of social analysis and policy.[3] But travel to the moon was itself once a staple of science fiction in the nineteenth and twentieth centuries.[4] Yet because of a single government program, exploration of the moon is now a historical event of more than forty years' standing. Technology has likely been accelerating since then, so events that seem even a greater stretch of imagination today may be the staple of the news of tomorrow. And unlike a lunar landing, even incremental advances in AI have direct implications for social governance.

Strong AI entails the creation of machines with the general human capacity for abstract thought and problem solving. It is generally conceded that if such machines were possible, they would soon surpass human cognitive abilities, because the same processes that gave rise to them could rapidly improve them. The machines themselves could aid in this process with their greater-than-human capacity to share information among themselves.[5]

The success of strong AI depends on the truth of three premises. The first is functionalism. Functionalism turns on the proposition that cognition is separate from the system in which cognition is realized.[6] Thus, abstract thinking can be equally realized in a biological system like the brain or in an electronic one like a computer. Under this hypothesis a system of symbols, when properly actualized by a physical process, regardless of whether that process is biologically based, is "capable of intelligent action."[7]

The philosopher John Searle is most prominent among scholars who challenge the notion that a machine manipulating abstract symbols can become the equivalent of a human mind. Searle provides the analogy of a Chinese room.[8] If someone is put in a room and asked questions in Chinese, he can be given written directions on how to manipulate Chinese characters so as to give answers to the questions in Chinese.[9] Yet because he himself understands nothing of Chinese, this manipulation of symbols is a poor simulacrum of human understanding.[10]

One powerful objection to Searle's analogy is that the entire system—the written directions plus the human manipulator—does understand Chinese.[11] Searle thus unfairly anthropomorphizes the subject of understanding. Confusing the proposition that AI may soon gain human capabilities with the proposition that AI may soon partake of human nature is the single greatest systemic mistake in thinking about computational intelligence—an error that science fiction itself has encouraged and perpetuated. But regardless of whether Searle is wrong or right about the conceptual possibility of strong AI, greater machine intelligence could still help humans in managing information and even formulating hypotheses about the social world. The metamorphosis of AI into a conscious agent may bear on the capacity of humans to control it. But continued progress in AI can bring very substantial social benefits even if the consciousness barrier is never reached.

The second premise undergirding strong AI is that computers will have the hardware capacity to mimic human thought. As described in chapter 2, raw computer power has been growing exponentially according to Moore's law. Assuming the computational capacity of computers continues to grow as Moore's law predicts, the hardware capacity of a computer is likely to achieve equality with a human brain between 2025 and 2030.[12] Even if this pace does not continue, it seems hard to believe that such capacity will not be reached by the midpoint of this century.

The third premise is that programmers will be able to provide the software to convert the gains in hardware into advances in AI. I have already suggested that software is also making very substantial improvements over time. Nevertheless, capturing the fluidity of human intelligence presents the greatest challenge.[13] In fact, some have argued that despite the previous growth in computational capacity and progress in software, AI has been largely a failure with little to show for fifty years of work.[14] This assessment seems far too harsh. For over a decade we have seen how computers have been able to defeat the greatest chess players in the world.[15] In 1997 IBM built Deep Blue, the machine that defeated world chess champion Garry Kasparov. While Deep Blue was a supercomputer, programs run on personal computers can now beat the best players in the world. Thus, what was once the frontier of AI has become the stuff of everyday computing.

This kind of progress on complex games is continuing. Although the rules of the famous Chinese game Go are simple and few, the nature of the game and the size of the board require any AI using the same computational strategy as Deep Blue to make approximately 1 million to the 10th power assessments of the game as it progresses. If each calculation lasted one microsecond, Deep Blue, using 1997 AI to play Go, would still require more time per turn than the age of the universe.[16] Yet researchers predict that ten more years of exponential growth in computer hardware and progress in computer software will make it possible to program an AI capable of calculating Go positions as thoroughly as Deep Blue analyzed chess moves.[17] Thus, AI is continuing to benefit not only from increased computational power but also from better ideas on how to make use of that power.

Of course, it is true that chess and Go are completely formal systems, mimicking only a narrow sliver of human capability. But it is hardly surprising that AI progresses from success in more formal and predictable environments to success in more informal and fluid ones. Software progress in tandem with the growing hardware capability continues to extend the kind of intelligence that AI can present.[18] A recent example is IBM's progress from developing Deep Blue in 1997 to developing Watson in 2011. This supercomputer specializes in playing *Jeopardy!*—a game of general knowledge. By 2011 Watson succeeded in beating the best *Jeopardy!* champions of all time.[19] This victory required the computer to react to natural language and to disentangle humor, recognize puns, and resolve ambiguity. In short, winning at *Jeopardy!* required Watson to operate in a much less formal world, one more like the chaotic world that human intelligence confronts on a daily basis.

While Watson used multiple formal rules in the form of algorithms to navigate this world, it is precisely its use of formal systems to compete in the world of more fluid human intelligence that makes its achievement a greater breakthrough than that of Deep Blue. Spinoffs from Watson will have implications for tasks of more general relevance to society than games. Already IBM is moving to sell Watson-like programs to help in medical diagnosis. These developments suggest that progress in AI falling short of strong AI will have large social effects over time. Some advances, such as a medical diagnosis program, are likely to complement a job done by a human, boosting productivity. But under other circumstances these programs could also substitute for a growing number of routine white-collar jobs. In my own field of law, the routine drafting of wills and contracts may soon be largely done by computers.

Continuing exponential growth in computational power is on the cusp of changing other key elements of social life. Cars run by computers now can autonomously navigate city traffic, another capacity that requires the ability to adapt to more fluid situations.[20] In fact, Google is confident enough that such systems can become routine in the near future that it has successfully petitioned the state of Nevada to make it legal for such autonomously driving cars to be on the road.[21]

Google's own core business—its search capacity—is itself an astonishing example of creating an intellectual function that far surpasses human capacities, and its search function is becoming more powerful. In part, the increased scope for search is simply a function of the increase in mobile computing. Mobile search opens up new search categories, like navigation and finding nearby services. But search will deepen as well as broaden in scope. It is deepening by the increasing capacity of the Web to categorize data. Search will aggregate the opinions of others from social media, adding to the capacity of our personal decision making. In the future the search function will use evidence of our past searches to help put queries into the context of our larger objectives. The mechanisms of search thus gain a greater underlying intelligence, morphing into agents that can actively help us organize our lives.

Other human-like abilities in which computers are improving rapidly include translating languages in real time and flying planes. A new program can write newspaper articles, and its architects even predict that it will garner prizes for journalism in five years.[22] The growing list of examples of specific intelligences generated by AI underscores its progress.

While these feats are impressive and exceedingly useful emulations of human behavior, they do not yet approximate strong AI, a general intelligence that spans the many functions and capabilities of human intelligence, to say nothing of a self-conscious intelligence. Recent trends in AI have turned away from direct attempts to mimic human intelligence, focusing instead on the comparative advantage of computers in turning their enormous calculating power to analyzing huge masses of data.[23] Through the use of multiple algorithms, such systems can accomplish feats that surpass human intelligence without in any way replicating its mechanisms.[24]

Some theorists, however, have suggested that this dependence on computers' brute strength in calculating capacity underscores the large gap between what has been accomplished and what needs to be achieved for strong AI. One critique is that AI must be embodied to become strong, and thus research into robotics is an important component of progress. Other researchers suggest that emotion will become a necessary component of strong AI, particularly any AI that aspires to consciousness.

Because of the focus on such obviously human qualities as embodiment and emotion, as opposed to simply computational capacity, some AI researchers are looking to mimic aspects of the human brain. They will be aided in this project by mapping the human brain, permitting a kind of reverse engineering. In 2007 IBM simulated half of a mouse brain.[25] While we perceive the gap between mouse and humans as very wide, in evolutionary terms these animals are relatively close and share a computational architecture. Moreover, in a world of technological acceleration, it is a fundamental mistake to view progress as linear.[26] As a result of exponential computational power, far more progress can be expected on brain simulations in the next ten years than in the previous ten and even more in the ten years after that. This progress is likely to open up new lines of AI research.

But the point here is not to prove that AI will succeed in replicating and then in surpassing human intelligence; it is just to suggest that such a prospect is plausible. In any event, greater progress in useful artificial intelligence can be expected. Even if AI does not actually exceed human intelligence, it may still offer useful insights that advance our collective decision making.

## The Threats of AI

AI has become a subject of major media interest. In May 2010 the *New York Times* devoted an article to the prospect of the day when AI will equal and then surpass human intelligence.[27] The article speculated on the dangers that such developments in strong AI might create.[28] Then in July 2010 the *Times* discussed computer-driven warfare. Various experts expressed concern about the growing power of computers, particularly as they become the basis for new weapons like the Predator, a robotic drone that the United States now uses to kill terrorists.[29] These articles highlight the twin fears about growing machine intelligence: the existential dread of machines that become uncontrollable by humans and the political anxiety about machines' destructive power on a revolutionized battlefield.

Bill Joy, former chief technologist for Sun Microsystems, is most eloquent on the first concern. He does not disagree with Ray Kurzweil that we are entering an age of unprecedented technological acceleration in which AI will become vastly more powerful than it is today.[30] But he does not share the view that this development will lead to technological utopia. In his article "Why the Future Doesn't Need Us," he raises the alarm that human beings cannot ultimately control these machines.[31] The power of his critique lies precisely in his acknowledgment of the wealth of potential benefits from strong AI. But for Joy, however great the benefits of AI might be, the risk of losing control of the intelligence created is still greater. In his view, man resembles the sorcerer's apprentice—too weak and too ignorant to master the master machines. This concern represents the culmination of a particular kind of fear that goes back to the Romantic era and was first represented by Dr. Frankenstein's monster, who symbolized the idea that "all scientific progress is really a disguised form of destruction."[32]

Fears of artificial intelligence on the battlefield may be an even more immediate concern. Nations have always attempted to use technological innovation to gain advantages in warfare.[33] Computational advancement today is essential to national defense. The Defense Advanced Research Projects Agency spends billions of dollars developing more advanced military mechanisms that depend on increasingly more substantial computational capacity.[34]

It is hard to overstate the extent to which advances in robotics, itself driven by AI, are transforming the U.S. military. During the Afghanistan and Iraq wars, more and more Unmanned Aerial Vehicles (UAVs) of different kinds have been used. Ten unmanned Predators were in use in 2001, and at the end of 2007 there were 180.[35] Unmanned aircraft, which depend on substantial computational capacity, form an important part of our military and may prove to be the majority of aircraft by 2020.[36] Under President Obama, the campaign against al-Qaeda is being conducted principally by drone strikers. It has been said that the F-35 now being built is the American military's last manned fighter plane and that the last American fighter pilot has already been born.[37]

Even below the earth, robots perform important tasks such as mine removal.[38] And already in development are robots that would wield lasers as a kind of special infantryman focused on killing snipers.[39] Others will act as paramedics.[40] It is not an exaggeration to predict that war twenty or twenty-five years from now may be fought predominantly by robots. The AI-driven battlefield gives rise to a set of fears that is different from those raised by the potential autonomy of AI. Here the concern is that human malevolence will lead to these increasingly capable machines wreaking more and more havoc and destruction.

## The Futility of the Relinquishing AI or Prohibiting Battlefield Robots

Joy argues for "relinquishment"—that is, the abandonment of technologies that can lead to strong AI. Those who are concerned about the use of AI technology on the battlefield focus more specifically on prohibiting or regulating weapons powered by AI. But whether the objective is relinquishment or the constraint of new weaponry, any such program must be translated into a specific set of legal prohibitions. These prohibitions are certain to be ineffective. Nations are unlikely to unilaterally relinquish either the technology behind accelerating computational power or the research to further accelerate that technology.

Indeed, were the United States to relinquish such technology, the whole world would be the loser. The United States is a flourishing commercial republic that can uniquely supply the public goods of global peace and security. Because it gains a greater share of the peace dividend than other nations, it has incentives to shoulder the burdens to maintain a global peace that benefits not only the United States but the rest of the world as well.[41] By relinquishing the power of AI, the United States would encourage rogue nations to develop it.

Thus, the only realistic alternative to unilateral relinquishment would be a global agreement for relinquishing or regulating AI-driven weaponry. But such an agreement would face the same obstacles as nuclear disarmament. As recent events with Iran and North Korea demonstrate,[42] it seems difficult, if not impossible, for such pariah nations to relinquish nuclear arms, because these weapons are a source for their geopolitical strength and prestige. Moreover, verifying any prohibition on their preparation and production is a task beyond the capability of international institutions.

The verification problems are far greater with respect to the technologies relating to artificial intelligence. Relatively few technologies are involved in building a nuclear bomb, but there are many routes to progress in AI. Moreover, building a nuclear bomb requires substantial infrastructure,[43] whereas AI research can be done in a garage. Constructing a nuclear bomb requires very substantial resources beyond the capacity of most groups other than nation-states.[44] AI research is done by institutions that are no richer than colleges and perhaps would require even less substantial resources.

Bill Joy recognizes the difficulties of relinquishment but offers no plausible means of achieving it. He suggests that computer scientists and engineers take a kind of Hippocratic oath that they will not engage in research with the potential to lead to AI that can displace the human race.[45] But many scientists would likely refuse to take the oath, because they would not agree with Joy's projections. Assuming some took the oath, many governments would not likely permit their scientists to respect it because of the importance of computational advances to national defense. Even without prompting from the government, many researchers would likely disregard the oath because of the substantial payoffs for advances in this area from private industry. All would have difficulty complying with such a directive because they could not easily predict what discoveries will propel AI forward in the long run.

For these reasons, prohibiting or substantially regulating research into AI is a nonstarter. Indeed, the relative ease of performing artificial intelligence research suggests that, at least at current levels of technology, it would be difficult for a nation to enforce a prohibition on research directed wholly against its own residents. Even a domestic prohibition would run up against the substantial incentives to pursue AI research, because the resulting inventions can provide lucrative applications across a wider range of areas than can research into nuclear weapons.

## Exaggerated Fears of AI

The threats from strong AI—both the fear that it represents an existential threat to humanity and the fear that it will lead to greater loss of life in war—have been exaggerated because they rest on conceptual and empirical confusions. The existential fear is based on the mistaken notion that strong artificial intelligence will necessarily reflect human malevolence. The military fear rests on the mistaken notion that computer-driven weaponry will necessarily worsen, rather than temper, human malevolence.

## The Existential Threat

The existential threat can be dissolved if there is a substantial possibility of constructing Friendly AI.[46] As defined earlier, Friendly AI is artificial intelligence that will not use its autonomy to become a danger to mankind. The argument for Friendly AI begins by rejecting the proposition that advanced artificial intelligence will necessarily have the kind of willpower that could drive it to replace humanity. The basic error in such thinking is the tendency to anthropomorphize AI.[47] Humans, like other animals, are genetically programmed in many instances to regard their welfare (and those of their relatives) as more important than the welfare of any other living thing.[48] But the reason for this motivation lies in the history of evolution: those animals that put their own welfare first were more likely to succeed in distributing their genes to subsequent generations.[49] Artificial intelligence will not be the direct product of biological evolution nor necessarily of any process resembling it. Thus, it is a mistake to think of AI as necessarily having the all-too-human qualities that seek to evade constraints and take power.

This is not to say that one cannot imagine strong AI capable of malevolence. One way to create AI, for instance, may be to replicate some aspects of an evolutionary process so that versions of AI progress by defeating other versions, a kind of tournament of creation. Such a process would be more likely to give rise to existential threats. Further, one cannot rule out that malevolence, or at a least a will to power, could be an emergent property of a particular line of AI research.

It is also true that humans may be able to enhance themselves by fusing themselves to AI devices to improve their intelligence and other capabilities. Such cyborgs might well have the will to power. But it is likely easier to regulate the acquisition of dangerous capabilities embedded in humans than to regulate research into AI in general. Monitoring AI research in general seems a hopeless task; monitoring individual citizens for such enhanced capabilities is more plausible.

Even a non-anthropomorphic human intelligence could pose threats to mankind. The greatest problem is that such artificial intelligence may be indifferent to human welfare.[50] Thus, for instance, unless otherwise programmed, AI could solve problems in ways that harm humans. For instance, computers with autonomy may naturally try to use resources efficiently and pursue their own objectives to the maximum. Such behavior could harm humanity, because an unrelenting focus on efficiency and maximal pursuit of any objective could easily bump up against human values other than efficiency and those objectives.[51] The solution likely lies in assuring that AI develops in an environment that cultivates the objectives of helpfulness and care.

In any event, the dangers from any variety of unfriendly or even indifferent AI provide strong reasons to develop a program of government support for Friendly AI. If Friendly AI maintains a head start in calculating power, its computational capacity can help discover ways to prevent the possible dangers that could emerge from other kinds of artificial intelligence. To be sure, this approach is not a guaranteed route to success, but it seems much more fruitful and practicable than relinquishment.

The question of how to support Friendly AI is a subtle one. The government lacks the knowledge to issue a set of clear requirements that a Friendly AI project would have to fulfill. It also lacks a sufficiently clear definition of what the end state of a Friendly AI looks like. This ignorance may inhibit establishing a prize for reaching Friendly AI or even any intermediate objective that makes progress toward this ultimate goal.[52]

The best way to support Friendly AI may be instead to treat it as a research project, like those funded by the National Institutes of Health. Peer review panels of computer and cognitive scientists would sift through projects and choose those designed both to advance AI and to assure that such advances would be accompanied by appropriate safeguards.[53] At first such a program should be quite modest and inexpensive. Once shown to actually advance the goals of Friendly AI, the program could be expanded—if necessary to the scale of the Manhattan Project during World War II.[54]

### The Concern about Battlefield AI

The concern about robots on the battlefield is likely misplaced, because there are three ways that the movement to robotic forces could be beneficial. First, robots render conventional forces more effective and less vulnerable to certain weapons of mass destruction, like chemical and biological weapons. Rebalancing the world to make such weapons less effective must be counted as a benefit.

Second, the use of robots should reshape the law of war in a way that saves human lives. One of the reasons that conventional armies deploy lethal force is to protect their human soldiers against death or serious injury. When only robots are at stake in a battle, a nation is more likely to use nonlethal force, such as stun guns. In fact, the United States is considering outfitting some of its robotic forces with nonlethal weaponry. The law of war's prohibitions against intentionally inflicting damage on civilians can be tightened in a world where the combatants are robots. In the long run, robots, whether autonomous or not, will be able to better discriminate among targets than among other kinds of weapons. As a result, the law of war can impose an effectively higher standard on avoiding collateral damage to humans and property that robotic forces would have to meet.[55] The need to protect robots from injury is less urgent than the need to protect humans, particularly as robots will, like other computational devices, become less expensive to make. Thus, the force authorized under international law for protection of robotic weaponry should be proportionately less as well.

Third, advanced computerized weaponry benefits the developed world, particularly the United States, because of its highly developed capability in technological innovation. Robotic weapons have been among the most successful in the fight against al-Qaeda and other groups waging asymmetrical warfare against the United States. The Predator has been successfully targeting terrorists throughout Afghanistan and Pakistan, and more technologically advanced versions are being rapidly developed. Moreover, the Predator is able to find its targets without the need to launch large-scale wars to hold territory—a process that would almost certainly result in more collateral damage.[56] If the United States is generally the best enforcer of rules of conduct that make for a peaceful and prosperous world, it is beneficial to the world that weaponry powered by AI will dominate the battlefield, because that is an area in which the United States is likely to retain an advantage. Thus, for the United States to relinquish computer-driven weaponry would be a grave mistake.[57]

It might be thought that any military exploitation of greater computational capacity is in tension with the development of Friendly AI. But the destructive powers unleashed by computer-driven weaponry do not necessarily entail the creation of strong AI that would lead to computers displacing humanity as a whole. Military activity by culture and design strongly reflects a system of hierarchy and subordination. Thus, nations will focus on integrating robots into their systems of command and control, putting an emphasis on assuring that orders are obeyed and making it unlikely that research by the military will be the origins of a strong AI that would seek to displace humans.

In any event, the United States and other advanced industrial nations can be better trusted than less developed nations to take account of these dangers, particularly if they have an ongoing Friendly AI program. The combination of support for civilian research into Friendly AI and continued deployment of computerized weaponry by the United States remains a better policy than the alternatives of relinquishment or robotic disarmament.

## The Benefits of AI in an Age of Accelerating Technology

Artificial intelligence differs from other accelerating technologies like biotechnology and nanotechnology in that it may help social governance by more successfully evaluating the consequences of various policies, including those toward AI itself. Insofar as artificial intelligence remains beneficent, it will facilitate the gathering and analysis of information that helps collective decision making about other technologies as well as about more quotidian matters of social policy, from welfare reform to tax policy.

In particular, improving AI will help with three important tasks in addressing social policy. First, it will help gather and organize the burgeoning amount of socially relevant data that technology continues to generate. Second, it will help form hypotheses about the data—hypotheses that human researchers may not come up with on their own. Finally, it will help improve our knowledge of social consequences by running simulations, and testing whether explanations of the effects of social policy are robust when small changes are made in conditions or assumptions.

Accelerating computer power is creating the phenomenon of "big data" as ubiquitous monitors track both material events, like the temperature at discrete localities or the paths of automobiles, and social events, like blog posts and tweets. But organizing such data will be an increasing challenge in a world of technological acceleration. The accelerating accumulation of information in the world makes mechanisms drawn from AI all the more necessary in order to sort and analyze that information.[58] Already the U.S. military is having trouble analyzing all the information acquired from its drones and other surveillance devices, because it lacks sufficient sorting and analytic capacity.[59]

The military's data problem captures future dilemmas for social decision making. As accelerating technology creates new complexity more rapidly in areas like nanotechnology, biotechnology, and robotics, social decision making must struggle to keep up with analyzing the wealth of new data. Societies prosper if they can use all the information available to make the best decisions possible. The problem now is that the information available to be processed may be swelling beyond our human capacity to achieve sound social decision making without the aid of AI.

The amount of data generated will in turn provide the basis for more conclusions about the effects of various social policies. Progress in AI can also be very useful by allowing computers to assist humans in explaining social phenomena and predicting the trajectory and effects of social trends. By 2020 computers are expected to be able to generate testable hypotheses. We will then no longer have to depend on the ingenuity of human researchers alone for formulating the full range of explanatory causes of social behavior and the effects of government policy.[60]

One way to understand this development is to see the increase in computational power as allowing more hypotheses to emerge from data rather than being imposed on the data. Greater computational power may allow computers to create competition between such emerging hypotheses, with the winner being the one that is objectively best supported. Ian Ayres, in his book *Supercrunchers*, has noted, "Trolling through databases can reveal underlying causes that traditional experts never considered."[61] In the future, however, computers may do the trolling autonomously, even constructing databases that will be optimally trolled.

Experts will also integrate their skills with computers, improving predictions within their fields. As Garry Kasparov has noted, humans plus computers remain better at chess than computers alone, even if the computer on its own is more powerful than any person on his own.[62] Human skills complement those of computers in hypothesis creation as well. Intuition and common sense help balance the computer's relentless application of formal algorithms.

The process of integrating human expertise and AI power may well be a recursive one. Different experts will use AI programs to look at the data and try to perfect models of an issue, like climate change. Then computers can use algorithms that discount certain experts because of past mistakes and weigh the combination of their predictions to form a result that is likely better than any single prediction.[63] This combined prediction then becomes grist for human evaluation through information markets. Similarly, regulatory experiments can begin through randomized trials, but computer algorithms can then tell us the rate at which to move toward the more successful regulation and ultimately when to end the experiment and choose the best regulation. The process is like using an algorithm to direct how money should be placed in different slot machines with unknown payout formulas. As different payouts are observed, it becomes rational to change the proportions of the slot machines on which bets are placed. The more general point here is that synthesizing a variety of methods—experiments, the analysis of experts, AI assistance, and prediction markets—may often yield better predictions than a single method can provide. Government can thus improve policy analysis by encouraging the applications of many different kinds of information technology.

Computer simulations will also become more powerful, permitting researchers to vary certain data from what exists and see what results.[64] Simulations will help enhance the robustness of modeling and empiricism that test the likely effects of social policy. Almost all social science analysis is complicated by statistical noise. Such noise can be the result of errors in measurements of the behavior to be explained or the result of the difficulty of capturing all the causes of human behavior. Simulation allows the researcher to vary some of the conditions and see if the results stay relatively similar. If they do, the empirical results are more robust and more likely to provide a map of the actual world.

One should not object to using simulations because of the fact that they are not themselves direct representations. Since measurement and other knowledge are limited, all data are themselves approximations of the world rather than direct representations. Computer simulations, when used in conjunction with empirical methods, are likely to make our approximations for social science more accurate and our conclusions less likely to be changed by the perturbations of the actual world.

A particularly promising area for simulations is so-called agent-based modeling, which allows a researcher to specify the behavior of virtual actors or agents.[65] These agents can then interact with another agent through repeated rounds of action and can then change their behavior through the rounds depending on what they have learned. These simulations permit researchers to find "emergent properties, that is properties arising from the interactions of the agents that cannot be deduced simply by aggregating the properties of the agents. When the interaction of the agents is contingent on past experiences . . . [agent-based modeling] might be the only practical method" of capturing a world with social learning.[66] Such modeling permits us to use the best evidence of regularities of human behavior and to refine our predictions by recognizing that actors may learn and thus change their behavior. In this sense, agent-based modeling has some of the advantages of prediction markets, which consider empirical analysis of past behavior but also force bettors to consider whether learning from past behavior will change future behavior.

Considered at their most general, these latter uses of AI expand our knowledge of the world by looking at the adjacent possible worlds, thereby helping society to see more clearly the likely results of government interventions. Increasing the capacity of AI to formulate hypotheses and to structure simulations could potentially render better decisions in the high stakes created by accelerating technology. That prospect itself justifies government subsidization of Friendly AI.

The greater capacity to process information should also better predict natural catastrophes, providing us with advance warning to prevent them or take preemptive measures to avoid their worst consequences. The more sophisticated the simulations and modeling of earthquakes, weather, and asteroids, and the better aggregation of massive amounts of data on those phenomena, the more accurate such measures are likely to be.[67] Moreover, by estimating the risks of various catastrophes, society is better able to make collective decisions to use its limited resources to focus on the most serious ones.

To be clear, AI's social utility does not depend on predicting the future with complete precision. Given the randomness inherent in our world, that feat is impossible, no matter how great the increase in intelligence.[68] Even if AI only makes clear the possibility of unexpected future contingencies, offers some solutions that would not otherwise be contemplated, and provides some assessment of their likelihood of success, it will help improve policy.

Government policy toward AI should be rooted in two complementary rationales: the need to deploy AI as an aid in collective decision making and the need to forestall dangers from strong AI. Each is individually sufficient to justify a program of support for Friendly AI. Together they make a most compelling case. The question of what degree and form of support is warranted will be subtle and difficult. But that is the right question to ask, not whether we should retard AI's development with complex regulations or, still worse, relinquish it.

# CHAPTER SEVEN

# Regulation in an Age of Technological Acceleration

MODERN GOVERNMENT is largely administrative government. Congress, by legislation, delegates substantial power to executive agencies. These agencies then promulgate regulations on a wide variety of subjects, from pollution to banking, from consumer safety to pharmaceuticals. While Congress oversees and influences the content of these regulations by conducting hearings on agency performance, it rarely overturns them. Courts also defer to the decisions of agencies and overturn only those regulations that are outside the scope of Congress's delegation or are not supported by evidence.

Administrative government though agency regulation was itself a response to technological change. The rise of the administrative state in the late nineteenth and early twentieth centuries coincided with the faster change brought about by industrialization.[1] Legislators were thought to have inadequate sources of information to use in formulating responses to such change. Indeed, some architects of the new legal and political order were aware that technological change was a cause of the administrative state's necessity.[2]

The rise of the administrative state can thus be understood in terms of restructuring government to make policy that is better informed about consequences. Under this rationale, Congress sets the basic policy objectives for an administrative agency, such as the Environmental Protection Agency or the Federal Communications Commission. But the agency is presumed to enjoy the comparative advantage of specialized expertise allowing it to access information and to assess whether a regulation would actually meet those objectives. The administrative state represents a first attempt to separate the distillation of preferences (done by Congress) and the prediction of consequences (done by agencies). This separation provides a good reason to consider how the function of an administrative agency can be updated with better information-eliciting rules.

More recently, new frameworks have tried to improve the assessment of consequences. In 1980 Congress established the Office of Information and Regulatory Affairs (OIRA) within the Office of Management and

Budget (OMB) to review regulations. President Reagan then issued an executive order establishing the regulatory review process by which OIRA generally assesses the costs and benefits of regulations and authorizes the promulgation of regulations only when the benefits outweigh the costs. This executive order on regulatory review creates a careful and reticulated structure for evaluating the consequences of agency action.

Every subsequent president has continued the regulatory review order with only slight modifications. Democratic presidents have focused a bit more on distributional consequences of regulation than Republicans have, but the basic cost-benefit framework has remained unchanged. The development of regulatory review is itself an excellent example of the ideal of consequentialist democracy in action.

In actual practice, however, the regulatory review process has been heavily criticized. One concern is that special interests with the most at stake in an agency's decisions are able to capture its regulatory process and distort regulation to ensure that consequences of regulation benefit the special interests rather than society as a whole. Thus, banking regulations often aid bankers, and regulations issued by the Labor Department help unions. A second concern is that the experts in the agencies will themselves be biased toward particular outcomes and imbued with an insular overconfidence in top-down decision making at the expense of more dispersed sources of wisdom. So, while the ideal is that an expert agency will make policy based on whether the consequences meet the objectives of Congress and the president—objectives that are themselves supposed to reflect the consensus of the American people—the reality is often otherwise.

Today's information technology has the potential to improve administrative government by changing the nature of the information on which it relies. Theorists in the New Deal may have been correct that Congress could not effectively gather information to create the detailed policy directives that would achieve its legislative objectives. But regulatory agencies have not proved all that much better. Only by requiring agencies to rely on the more dispersed sources of information provided by markets and experiments is the situation likely to improve. By creating more objective measures of results, the use of such information technologies also constrains the influence of special interests. More generally, these new mechanisms empower encompassing interests, because they make clearer to the public which policy instruments are likely to achieve the goals established by law, replacing ideological frolics with more fact-based analysis.

Faster technological change will also challenge the administrative state in other ways. Accelerating technology not only expands the number of regulatory issues by creating more technologies with potentially dangerous spillovers, like pollution, but also creates the need for new kinds of crosscutting analysis. For instance, acceleration in technology means that current regulatory policy may be affected by the dramatically different technology that could be available in the near future. Administrative government must systematically project future technology to formulate current policy.

## Decentralizing Centralized Regulation

It is difficult for agencies to acquire the information to determine whether a regulation will achieve the objectives set by Congress. Policy effects are often disputed and uncertain. The uncertainty is exacerbated because regulations have secondary effects that may undermine the achievement of the very objectives Congress has established. For instance, assume that Congress has provided that the Federal Aviation Administration should impose safety requirements on airplanes when the benefits of a regulation exceed its costs. Even assuming further that the agency is able to set a value to lives saved and injuries prevented, it is often unclear what the effect of the regulation will be. How many accidents will it prevent? Will its costs raise airline fares and encourage people to take less expensive but more dangerous forms of transportation, such as cars? Will it discourage innovations that would lead to safer airlines in the long run? Or assume that the Federal Communications Commission is told that its licensing decisions should ensure that a diversity of voices is heard by the American public. Will rules against media concentration advance interests of diversity? Or will they discourage the kind of innovation that big companies can best undertake, thus preventing new programming that would increase the variety available to consumers?

Administrative law attempts to make agencies more informed by requiring them to hold a period of notice and comment in order to permit the public to assess the regulation. But this process has obvious limitations. The public has little incentive to respond, and special interests may skew the administrative record with self-serving analysis. It is thus not surprising that agencies continue to rely principally on the assessments of their own experts. As a result, they engage in substantial top-down decision making without systematic access to more dispersed sources of knowledge. But these experts themselves are influenced by the interests of those they regulate, because these interests wield substantial control over their future employment prospects and over their reputation in a specialized area of expertise.

The new information technologies can substantially improve agency decisions by permitting them to tap into less insular and more dispersed sources of information. The scope for experimentation is potentially as large as the regulatory agenda of an agency. Whenever there is a plausible debate about which regulation will better achieve the goals of the agency, or whether no regulation at all is the best possible outcome, there is room for considering experimentation. Thus, an experiment in which certain regulations would be imposed on some factories and not on others offers the real prospect of determining whether those regulations are useful. The standing presidential order on regulatory review should be revised to direct agencies to consider such experiments to resolve disputed issues of policy. Bureaucratic inertia may resist a more relentlessly experimental approach, but the creation of a unit devoted to experimentation will create an impetus for change. In a bureaucracy, function follows form.

Government decentralization is also useful to create the conditions for empirical learning. But there has been a growing trend for agencies to preempt state action with uniform federal law. As a result, the standing executive order that promotes federalism should be revised to direct federal regulators to take into account the opportunity for empiricism afforded by federalism before preempting state regulatory regimes.[3] Through both of these initiatives, agencies would restructure themselves to be more self-consciously experimental.

Prediction markets offer the other most important technology for improving regulatory decisions.[4] Instead of offering the public the opportunity to simply comment on a proposed rule, a conditional market could be predicated on a few alternatives. The rule could then be assessed to see whether it will meet its purported objectives. For instance, will a safety rule actually save lives? A market could be made in lives lost in the industry conditional on a new regulation and conditional on its absence. How much will a regulation raise costs? Another market could be made in costs measured by a designated benchmark conditional on the introduction of the rule and conditional on its absence.

It is true that such prediction markets may raise more difficult questions than prediction markets on more general subjects. Even with the subsidies suggested in chapter 4, they may be subject to manipulation if they do not attract a sufficient number of participants. The problem of manipulation may be especially acute, given that regulations can dramatically affect the income and opportunities of certain industries. But again we must be careful of the nirvana fallacy. Agencies are regularly confronted with all sorts of special interests that make claims about the consequences of regulations. The advantage of prediction markets is that they provide incentives to foretell accurate consequences. Even in settings rife with special interests, such markets have the potential to generate more accurate information about consequences than the notice and comment process or other procedures that do not reward participants for correct assessments.

Nevertheless, regulatory prediction markets may well need to be designed differently than prediction markets on more general policy. For instance, there may be greater need for conflict-of-interest rules that bar parties with a direct interest in the regulation from participating in the market. The subsidies for such prediction markets may need to be greater to attract participation. In short, regulatory prediction markets are themselves appropriate *subjects* of experimentation. Indeed, under certain circumstances it may be plausible to create competing kinds of prediction markets in order to discover which kind is superior in its predictions. Experimenting with prediction markets underscores the synergies between empiricism and prediction markets. For this reason it is particularly important that OMB establish a unit focused on prediction markets and their design in various regulatory domains. These experiments should be undertaken now, because accelerating technology will raise the regulatory stakes in the future.

In the long run, tapping into the broader sources of information that empiricism and prediction markets facilitate will also give agencies more legitimacy, because they will help demonstrate in a transparent manner that the regulation embraced is actually likely to achieve its objectives. Greater legitimacy is important in an era of technological acceleration, because in order to protect against relatively small probabilities of catastrophes, agencies may also have to take quite radical action that will adversely affect members of the public. The public is more likely to support such initiatives when agencies are seen to be using the best possible sources of information.

## The Present Future of Regulation

Accelerating technology makes future technological change more relevant to current regulation. In a world of technological stasis, there would be no need to consider the path of future technology within the regulatory process, because future technology would not substantially modify the costs and benefits of contemporary regulation. But as technology changes faster, the stream of costs and benefits also can be transformed, making future technology increasingly central to *current* regulatory decisions. Thus, modeling the course of technological change needs to become a central and formal part of any cost-benefit analysis. Moreover, any government regulatory regime should consider eliciting technological improvement as a policy option. Subsidies and prizes for technological innovation should be made an integral part of the regulatory review process.

First, the acceleration of technology may affect the timing of regulation or its current content. For instance, assume we could predict the development of carbon-eating plants, a likelihood considered reasonable by the celebrated physicist Freeman Dyson. It is then possible that we would be less interested in controlling current emissions because some of the effects of these emissions might be reversible.[5] Accordingly, at times it may be more effective to delay certain kinds of emission standards because those kinds of emissions can be cleaned up at low cost in the relatively near future.

But the possibility of technological innovation will sometimes favor *more* current regulation rather than less. Climate change again provides a good example. Here technological innovation may ameliorate a central problem of regulating global emissions: the difficulty of achieving effective regulation without the participation of other carbon-emitting nations.[6] If regulating emissions domestically encourages quicker innovation, this new technology may be sufficiently inexpensive to be used by nations that did not participate in the regulatory regime. So the capacity of a regulation to elicit innovation may justify the regulation even in the absence of a global regulatory consensus.

Thus, the future shape of technology will become more and more important to current regulatory decisions. Some technological change can be predicted with relative certainty. Moore's law predicts the continuous and exponential expansion of computer capacities for the next two decades. As a result, scientists have sometimes postponed solving mathematical problems that can be resolved rapidly within a few years. Some have argued that other processes, such as the improvement of solar panels, are also now approximating a predictable glide path.[7]

Other technological change can be predicted through information markets of the kind already discussed. These markets can take expert claims about technological prospects and put them to a market test. The need to predict the future course of technology for current regulation provides yet another reason for establishing a prediction market unit within OMB.

Our views of the shape of future technology may shift over time. Thus, regulatory structures should be designed to update regulations without a laborious process. In its original regulation an agency could specify that rules should be modified if certain technological advances arise.[8] Building such revisions into regulation will also make it easier for companies affected by regulations to plan for the future.[9]

Accelerating technology also requires that we better integrate regulatory decisions with government decisions about encouraging technology. Even if technology does not solve a regulatory problem today, it may offer a full or partial solution in the future. Encouraging new technology may offer the most cost-efficient way of solving a social problem, either as a substitute for regulation or in conjunction with regulation.

It might be thought that the government should always promote general policies for encouraging innovation as opposed to any particular technology. According to this argument, if the government chooses the appropriate regulations, companies will have incentives to invent technology that will enable other companies to comply with those regulations most cheaply. Regulation in this sense would become the visible hand of technological progress.

This analysis does not necessarily work in the world of domestic and international politics, however. A principal difficulty is that special interests may inhibit the creation of purely regulatory solutions. Programs with concentrated costs and diffuse benefits are difficult to enact.[10] For instance, concentrated groups like energy companies may successfully oppose regulations to stop global warming because they are well organized. In contrast, government can use general revenues to subsidize technology or to establish a technology prize to solve the problem that regulation addresses. Instead of a tax on emissions, it could provide a prize for technology that dissipates emissions from the air in specified amounts. Programs with such diffuse benefits (many corporations may think they have a shot at the prize) and diffuse costs (all taxpayers pay for it) are easier to enact. In a world of technological stasis, often the regulatory solution may be the only alternative for solving a problem like pollution. However, in a world of technological acceleration, solutions through innovation may be not only technologically feasible but also politically optimal.

Another reason to believe that even efficient regulation will not stimulate the optimal amount of innovation in technology is the international nature of many regulatory problems. Again, climate change provides a salient example. If other nations, particularly developing nations, do not regulate emissions, companies operating in those nations will not have the proper incentives to buy technology to reduce emissions and companies will not have the proper incentives to produce that technology.[11] Moreover, because of the imperfect enforcement of intellectual property rights in much of the world, companies may fear that foreign nationals will appropriate their inventions. Thus, domestic regulation may not effectively curb global pollution, because it will neither provide sufficient incentives for innovation nor curb emissions abroad. Accordingly, prizes or subsidies for a particular technology may prove a better alternative.

When the government cannot easily observe the quality of the research, prizes may be superior to direct subsidies.[12] Technological acceleration may increase the utility of prizes, because it makes possible diverse approaches to create innovations, which may make it more difficult to observe research quality. Offering prizes has similar advantages for policy to those of federalism or random experimentation. All of these mechanisms seek to generate information that the centralized government cannot discover on its own. As with random experimentation, private organizations are showing government the rich possibilities of embracing a less top-down approach. For instance, the X Prize Foundation is offering prizes on such diverse subjects as providing faster ways to sequence the human genome and creating an affordable car that gets one hundred miles per gallon of gasoline.[13]

It would be wise for Congress to appropriate money for a prize fund and permit OMB to determine how it should be spent. Technology funding for regulatory purposes, like regulations themselves, requires time and expertise that Congress lacks. Although agencies may be more expert than Congress, it does not follow that they should make decisions on how to incentivize technological innovation on a top-down basis. Even in the decisions about where and how to offer support and incentives for technological innovation, prediction markets are likely to prove quite helpful.

## Intergenerational Equity

The future of accelerating technology raises the problem of intergenerational equity in a form that is more accute than previously seen. If it is recognized at some point that accelerating technology likely makes the next generations far wealthier than this generation, our current policies may need to take into greater account the possibility of the intergenerational redistribution.[14] Any substantial economic growth over time creates a potential problem of intergenerational equity. But accelerating technology makes the issue far more pressing, because it increases the gap in well-being between proximate generations, making redistribution tenable.

The consequentialist function of democracy takes no position on the appropriate degree of intergenerational redistribution. The appropriate degree, both within generations and among them, is determined by the preferences of individuals. But accelerating technology may change our view of the direction of the redistribution as a matter of fact and so change our assessment of appropriate policy, given our distributional preferences. For instance, Social Security has been criticized as a transfer from the young, who as a class have relatively few assets, to the old, who as a class have relatively many assets. Considered without regard to technological acceleration, Social Security thus may appear to fail the objective of creating greater economic equality. But to the degree that accelerating technology will substantially increase the well-being of the younger generation, the factual premise of this argument is undermined.[15] Technological acceleration raises more acutely questions about who is better able to bear the costs of regulation—the current generation or the generations to come.

Indeed, it is possible that in an era of accelerating technology inter-generational equity might come to be thought of as more pressing than intragenerational equity. Today an entire generation shares the benefits of technological advances as innovation now filters more rapidly down even to the nation's poor.[16] To give one example, the almost universal adoption of cell phones has happened much faster than the adoption of televisions.

It is true that accelerating technology may make wages within a generation more unequal by allowing the most able to amplify their earning power. But the innovative ideas at the root of today's technology can be used simultaneously by everyone. Even if expressions of these ideas are legally protected, their price tends to fall quickly as they are succeeded by even better inventions. Thus, there is a more effective equality of consumption, because almost everyone enjoys the fruits of so many discoveries. To return to the personal example offered in chapter 1, I can now download a calculator for free that would have cost two thousand dollars when it was first introduced. A generation with almost universal and rapid access to such technological innovation that affects a broader range of life experiences might have a greater measure of effective equality, even with large wage differentials. But the same accelerating pace of innovation makes the gaps among the wealth of succeeding generations even larger, as each generation will enjoy far more technological benefits than the last.

To evaluate issues of intergenerational equity, we need to predict the income and the well-being of future generations. Because of the difficulty of measuring changes in well-being in an era of technological accelera-tion, benchmarks marking income, life expectancy, and technological im-provement in future generations are all relevant. Prediction markets once again furnish the mechanism that can help with such information, be-cause they can provide a distribution of the likelihood of various results. This distribution can be used as a basis for assessing the most likely level of well-being of future generations.

However, the growth rates and innovation that will increase the well-being of future generations depend on current regulatory and other governmental decisions. One danger is that politicians will engage in so much intergenerational redistribution as to reduce the incentives of the young to innovate, thus depressing the wel-fare of all generations.[17] A constraint against this tendency may be the conditional prediction markets described above. Prediction markets can be made in rates of economic growth and longevity on important regulatory or tax schemes. These markets can provide information about the degree to which such redistribution would destroy wealth, harming the children and grandchildren of the current generation.

Thus, accelerating technology may create not only new innovations but also new versions of old questions—questions that are likely to be faced at first in the administrative context, given its already reticulated framework for assessing consequences, including distributional consequences.

## Accounting for Possible Catastrophes and Elysian Benefits

Another issue raised by accelerating technology is how to consider small probabilities of extreme consequences, both beneficial and catastrophic. Because technology is proceeding so rapidly across so many fronts, it creates more of these possibilities than ever before. Unlikely events do not have to be considered in a regulatory cost-benefit analysis if their effects are not too great, because the effect of an event must be discounted by its probability. In contrast, an event that will have a potentially huge effect must be considered even if its probability is relatively small, because the effect even when discounted by its probability can be large.

Richard Posner, the federal appellate judge and legal polymath, has considered how to integrate evaluating the problem of catastrophes into regulation. He usefully notes that although there may be a great amount of uncertainty in assessing the likelihood of various catastrophes, regulation can address uncertainty.[18] Even if regulators cannot pinpoint the probability of a catastrophe, they can suggest a range of probabilities for the likelihood of the catastrophe. It then becomes possible to provide a range of spending that might be considered reasonable in order to avoid the catastrophe. That range provides a metric to critique inadequate or excessive spending.

The extraordinary benefits of accelerating technology represent the flip side of potential catastrophes.[19] Accelerating technology increases the possibilities of both kinds of outcomes. Indeed, the same technological advances may raise the prospect of both. Nanotechnology may provide mechanisms for improving health, but it may also create new kinds of dangerous pollution. Thus, a question for regulatory review is how to compare the probabilities of catastrophe and of enormous, cascading benefits that might arise from an accelerating series of technological breakthroughs. Elysian benefits may sometimes be as relevant to regulatory calculation as the Stygian costs of doomsday scenarios.

One might argue that we should weight catastrophic loss more heavily than an equal probability of an extremely beneficial outcome, because the total catastrophe could preclude the possibility of any future life and the potential progress life brings with it. But this argument seems too facile in the case of technologies that will help us avoid a range of catastrophes.

We saw in the last chapter that artificial intelligence could have beneficial externalities in helping to predict and avoid catastrophic events. It can also help evaluate the hidden benefits and costs of other accelerating technologies. Thus, even a relatively small probability of creating strong AI might justify a substantial government program of subsidies or prizes designed to stimulate its creation because of the very substantial benefits it would bring, even given the risks that strong AI may pose to humanity.

Radical life extension is also an Elysian benefit that may depend on risky choices. Radical life extension is premised on the idea that innovations in medical technology may permit people to live longer until the next wave of medical innovations allows them to survive for yet another round of innovation.[20] If very long or even indefinite life spans are thought to be a future possibility, individuals could choose to seek such treatments even at the risk of sudden death. This choice presents a secular version of Pascal's wager.[21] Just as it may be rational to believe in God even if there is only a tiny probability of his existence, given the infinite benefits of such belief, it may be rational to take risks to live to a certain time in the future if by so doing one can enjoy what one regards as the almost infinite benefits of an indefinite life. Such considerations could be relevant to the regulatory analysis of approving pharmaceuticals. A drug might likely be harmful with only a small chance of extending life. Yet if medical innovations are expected to accelerate, even a small probability of surviving until those innovations occur might permit a patient to surf to a very long life.

Prediction markets and computational modeling might help regulators consider how much they should count possibilities of catastrophes and Elysian benefits in their calculations. One advantage of prediction markets is that they provide not only evidence of whether an event is unlikely but also evidence of the probability an event will happen even if it is unlikely. Regulatory choices about the structure of our health care system bear on the pace of medical innovation and thus on the possibility of radical life extension.[22] As prediction markets and computational modeling are refined, they may help assess the effects of health care policy on longevity, including on radical life extension.

Richard Posner has denigrated the utility of prediction markets for events that depend on scientific analysis on the grounds that these issues are the domain of scientists and that the man on the street can add little.[23] But even if it were true that only experts could contribute in complex social and technological problems, the bounds of expertise are unclear. As the example of the economist Simon betting against the biologist Ehrlich discussed in chapter 4 shows, many problems can be addressed by different disciplines with resulting wide differences of perspective. A market helps mediate which set of experts is more likely to be right. Moreover, the layman may not be exactly sure which expert is right, but he may have enough information about the relative insularity and biases of experts to perform a useful function. For instance, he may think economists face a more competitive market and more scrutiny from colleagues than do scientists from their colleagues on issues of public concern, and so he will tend to favor economists' predictions about resource depletion over those of climate scientists or geologists.

Prediction markets may also help regulators indirectly by increasing public awareness of the relevance of important events that are relatively unlikely to occur. These markets address a problem that people face in taking account of small probabilities. When the probability is sufficiently small, individuals seem to dismiss the event from their calculations, and they might not give enough weight to events that are less likely than not to happen.[24] But prediction markets can dramatize the probability of an event by providing an estimate of its liklihood and they can make individuals more aware of the relative odds of various events, even if they are unlikely. This comparison helps society apply its resources to the catastrophes that would cause most damage, as discounted by the probability of their occurrence.

Like other structures of government, the administrative state must become a better mechanism for social learning from dispersed sources of information. This kind of improvement is all the more important because the tasks that regulatory review must consider are broader in scope and may be more urgent today because of the high stakes of catastrophic risks and Elysian benefits. Using the new tools of information technology offers the best prospect both of reducing the errors inherent in bureaucratic planning and of making our accelerating future more relevant to current regulation.

# CHAPTER EIGHT

# Bias and Democracy

SOCIAL KNOWLEDGE IMPROVES collective decision making only if new information changes minds. But internal bias presents an important obstacle to updating on the basis of external evidence of the world. Indeed, some political scientists and psychologists believe bias is so pervasive that little if any updating takes place in politics. If bias prevents more information from modifying electoral outcomes, better deployment of information technology and better information-eliciting rules will not help solve problems of governance.

A host of biases infect political decision making. Nevertheless, even now these pervasive biases do not present an insuperable obstacle to democratic updating. The fundamental building blocks of a modern democratic system, from voting rules to the nature of representation, constrain bias by economizing on the number of people who need to update and by providing incentives to key players to temper their biases.

## The Nature of Bias

Bias represents a departure from rational updating on the evidence. People are biased in their daily lives even apart from politics, and psychologists have cataloged these biases in a systematic way. But whatever the departures from rational updating in daily life, it would be extravagant to deny that updating routinely takes place. It is an axiom of cognitive science that people often change their minds in the face of new information,[1] even if biases and cognitive limitations make such updating imperfect. Experiments suggest that individual decisions are often better when there is better access to information and facts.[2]

Evolution offers the most powerful proof that innate biases are not so substantial that they preclude updating. If humans did not rationally act on evidence, they would be less likely to make it to reproductive age.[3] Failing to update would make it likelier that one would miss the opportunity to trap an animal for one's meal or avoid the predator behind a tree.

Indeed, changing behavior on the basis of new information is not a trait limited to humans; other living things, even insects, possess it as well.[4]

It follows that information should help individuals update their political beliefs, unless there is something about collective political life that precludes updating. One fact about politics may make citizens more prone to bias in collective decision making than in decisions made individually. People's decisions in politics have less influence on their personal lives than do their decisions about jobs and family, because their individual decisions have little effect on political outcomes. The relative lack of personal accountability gives rise to a greater scope for bias. It is thus worth describing some of the most important biases.

## Varieties of Bias in Politics

Psychologists and political scientists have outlined a variety of biases that particularly affect politics. Some kinds come naturally to a minority, and others are more likely to be imposed by a majority. Some may be innate to the individual, and others are the product of rational collective action. But all make it harder for democracy to update on facts about the consequences of policy.

### Special Interest Bias

If a group of milk producers wants to gain price supports, it is biased toward arguments about the good social consequences of propping up milk prices. Given that special interests are well placed to exact benefits from the government, they are not likely to easily assimilate new information against their position, because it may interfere with large gains. Worse still, special interest groups have strong incentives not only to disregard information contrary to their interest but also to suppress it and, where necessary, to attempt to drown it out with noisy untruths that may confuse and bias others.

The ignorance of most voters often impedes the ability of democracy to see through the special pleading of special interests. Because they cannot gain benefits from the state through collective action, and because their individual votes are exceedingly unlikely to influence an election, most voters will not invest in a study of the issues and may be easily swayed by special interest disinformation campaigns. As a result, special interest misinformation can impede democratic updating.

### "Knowledge Falsification" by the Majority

Individuals may slant or falsify their information because of their interest in their own appearance or status rather than because of any interest in influencing policy. This kind of bias is ultimately more likely to stem from a majority's positions on an issue than a minority's position, because a majority can more easily enforce social conformity. Individuals may withhold or distort their assessments of the truth because they fear the social opprobrium from dissenting from what they believe is the majority's view. Economist Timur Kuran has analyzed this tendency as "knowledge falsification."[5]

For instance, assume that the public relations campaign of a teachers union has persuaded most people in a community that small classes promote student learning when the facts are otherwise. If a citizen or even an expert has correct information on the subject, he may slant or falsify that information because he does not want to suffer loss of popularity from opposing a policy with majority support. Insofar as the union has managed to link its factual claim to a widely shared objective, like helping children, it may be even harder for someone with knowledge at odds with that claim to inject his information into the public sphere. By espousing a view contrary to a widely held perception that a policy helps children, the skeptic will be seen not merely as wrongheaded but as an enemy of the people.

Knowledge falsification can lead to what are known as information cascades or bandwagons.[6] The more people support a position, the more social pressure they can bring to bear on possible dissenters not to question it. Knowledge falsification has the potential to sweep away contrary opinions and thus impede democratic updating.

### Innate Majoritarian Bias

Other biases of the majority may be natural—innate to the human psyche—rather than a product of cultural imposition. An example of such an enduring bias is xenophobia, or at least natural distrust of foreigners. This bias may explain in part the often intense opposition to free trade, despite its large economic benefits, and the widespread belief that a large part of the federal budget goes to foreign aid, despite its paltry share of federal expenditures.[7]

### Status Quo Bias

Another pervasive bias is attachment to the status quo. Of course, attachment to tried institutions at the expense of untried innovations is not necessarily a status quo bias. It may reflect a rational response to the constraints of information and human intellect that make it difficult to design better institutions than those that have been tested through the ages. That factual claim about the limits of our knowledge is the essence of conservatism of the kind made famous by Edmund Burke.

But psychologists suggest that innate tendencies of aversion to loss and attachment to things we possess—called the "endowment effect"—make the status quo bias more powerful than this rational response to human limitations. In particular, political scientists suggest that people become more entrenched in their partisan positions as they age, making them potentially less responsive.[8] Moreover, as people grow older they also lose incentives to invest in learning about new technologies and their effects, including the information technologies that are so important to modern political learning. They cannot gain as much from new knowledge, because they are not likely to live and work as long as the young.[9] Thus, they are less likely to update and more likely to stick with prior beliefs based on information from an earlier age.

Given that accelerating technologies are likely to require rapid changes in policy, status quo bias likely creates an increasingly significant constraint on updating. Self-driving cars may soon become not only possible but also safer than cars that humans drive. As progress continues, self-driving cars could be linked through a single network. At some point everyone on the road will likely be far safer if all cars become part of a self-driving network. And yet people who have been driving cars for a lifetime can be expected to doubt that reality and to oppose laws that would force them into these new vehicles.

### Cultural Cognition and Motivated Reasoning

Citizens may be biased not only for the instrumental reason of acquiring benefits from the government but also for the personal reason of affirming their own identity. Dan Kahan, Donald Braman, James Grimmelman, and a variety of co-authors have suggested that individuals are divided into different worldviews such as egalitarian, individualist, and hierarchical.[10] On this conception, citizens tend to regard new information through a particular mind-set and do not easily allow updated information to sink in. Because of what cognitive scientists have called "biased assimilation"—a species of confirmation bias—people tend to view evidence through their own cultural prism, dismissing it when it creates "cognitive dissonance" with their own worldview.[11] More generally, confirmation bias is a species of a larger category of motivated reasoning or motivated skepticism—"the systematic biasing of judgment in favor of one's immediately accessible feelings."[12] As well as confirmation bias, such motivated reasoning encompasses a tendency to seek out information that confirms one's prior beliefs and to spend more time undermining evidence inconsistent with those beliefs. Evidence for motivated reasoning has been found in laboratory experiments.[13]

Moreover, attachment to one's own views is likely to help insulate people from others' challenges to one's favored policy prescriptions. Because of "naïve realism," citizens are likely to regard their own assessments of policy as a product of an objective understanding of the world while regarding their opponents' assessments as a product of subjective and benighted worldviews.[14] As a result, they will not be influenced by people of different worldviews. Thus, in the normal course of events, individualists, who oppose gun control, tend to dismiss information that guns cause harm while egalitarians, who embrace gun control, tend to dismiss information that gun possession can save lives. More generally, Republicans dismiss positions held by Democrats, and vice versa. While such bias may not be instrumentally rational because it gains no pecuniary benefits, it does help protect the social status of such groups as well as maintain the sense of their own identity.

## Framing

Bias does not derive only from citizens. Politicians actively try to promote biases that work in their favor. They have every interest in imposing a favorable frame of reference into which the facts are to be fitted. This frame seeks to bias citizens' judgment about the relevance of new facts so that new information, whatever it is, increases rather than decreases their support from voters. According to the political scientist James Druckman, framing occurs when "in the course of describing an issue or event, a speaker's emphasis on a subset of potentially relevant considerations causes individuals to focus on these considerations when constructing their opinions."[15]

For instance, Democrats talk about government outlays as investments in an attempt to make people think about the possible benefits rather the costs of these outlays. Republicans talk about government spending and the taxes needed to support that spending to focus on the diminishment of individual resources. Some theorists, like George Lakoff, a psychologist at the University of California, Berkeley, argue that such frames can wholly blind voters to contrary evidence. In this extreme form, the framing claim seems implausible because it assumes that voters are almost infinitely gullible.[16] A politician who frames taxes as membership fees in a club, as Lakoff recommends to his fellow Democrats, is likely to garner more ridicule than votes. But James Druckman and his colleagues have done a series of experiments that suggest that framing does have substantial effects in biasing people, meaning they are likely to experience more difficulty in assimilating evidence that is inconsistent with the frameworks they are given.[17]

The existence of such biases poses a challenge to any naïve optimism that reducing information costs will necessarily lead to better social decision making. But a well-functioning democracy already has mechanisms that reduce, even if they do not eliminate, the effect of bias on updating on new information.

## Democratic Institutions and Cognitive Diversity as a Constraint on Bias

In the long run, democracy is often able to overcome biases within the citizenry, because it takes only a majority or a relatively modest supermajority of people to change ordinary legislative policy. Thus, if many, or even most, people are imprisoned by their own worldviews, misled by politicians' frames, or remain ignorant of all new information relevant to public policy, the shift of a relatively small portion of voters can often make a decisive difference. Some voters are swing voters, poised to switch easily. Other voters are partisans, but weakly so, and some of them might switch as well. Even the turnout of strong partisans may rise or fall, making a difference to the result. Thus, if new evidence can change the behavior of a subset of voters, it is worth trying to make use of the opportunities provided by new technology to muster better evidence and to distribute it more widely. We illustrate democratic constraints on bias through considering the federal system, but most states have some similar structures.

### Democratic Institutions

Both in deciding who the representatives of democracy are and in passing legislation, citizens and representatives in the middle can often make a decisive difference.[18] In elections for representatives that involve contested issues, individual voters in the middle—both swing voters and weak partisans—are less certain of what policy to pursue than those at the extremes.[19] To be sure, this effect of democracy can be muted by partisan gerrymandering.[20] However, gerrymandering is not a feature inherent in democracy but rather a contingent obstacle.

In legislatures as well a coalition can succeed in passing new legislation if moderates support it, even if legislators of more extreme views remain opposed.[21] It is true that in our national legislature the size of that coalition is further increased by the filibuster rule in the Senate, but even there legislation can pass with 40 percent of the senators in unbending opposition.[22] And on the important matter of the budget, the reconciliation process allows legislation to pass in the Senate with a bare majority.

In legislative politics, middle-of-the road legislators are sometimes constrained by their party attachments to vote for the party's positions. But American parties are relatively weak and legislators often vote independently.[23] Moreover, swing voters and weak partisans can wield substantial influence on whichever party is in power in the legislature. Thus, over time parties have some tendency to move toward the swing voters' preferred positions.[24]

The power of the middle is clearly at work in the election of the president. Presidential elections are generally won in swing states, where by definition independent and weak partisans remain the key to victory. Democrats cannot rely for victory on their partisans or egalitarians in general. Republicans cannot rely on their partisans and those who value liberty over equality. Candidates must appeal to the broad middle.

Indeed, the system that the framers of the Constitution established for election of the president helps temper some kinds of bias.[25] The median voter in any particular state is likely to be more influenced by parochial biases, such as special interests peculiar to their specialized industries, than by the median voter in the nation as a whole. The framers saw this filtration system as part of the advantage of the large republic.[26] According to James Madison, biases are filtered out in the large republic so that "a coalition of a majority of the whole society can seldom take place on any principles other than those of justice and the general good."[27] Madison was too sanguine about filtration. The swing states in the electoral college often have some parochial interests: think of Iowa and the support by both parties for ethanol, despite its dubious cost-effectiveness as a fuel. And democracy misfires more than he expected. But it is true that constitutional mechanisms of a continental republic impose some constraint on bias.

Less ideologically committed voters have yet another resort for enforcing their views in legislative and presidential elections. In the American system of bicameralism and presidential power, swing voters can split their votes to bring about divided government in which different parties control at least one House of Congress or the presidency.[28] This political dynamic forces compromise among the parties and provides the center with more power.[29] Divided government is the predominant reality of modern American government. By 2012, of the last sixty-two years, all but twenty-four will have been marked by divided government.[30]

To say the middle exerts substantial influence in American politics is not to claim that the middle's triumph is immediate or inevitable in every case. In electoral politics, winning an election requires a balance between turning out the base and appealing to the middle.[31] Sometimes turning out more extreme voters can have an effect. But even here information can make a difference. It is more difficult to turn out more extreme voters in a political climate where the policies supported by their leaders seem to have had bad consequences. In fact, the reduction in willingness to vote is a rational response to the new information; one is more likely to incur costs, such as the cost of voting, to express a belief in which one has more confidence than a belief in which one has less confidence.[32] Thus, new information likely has an effect on more extreme voters as well as swing voters and weaker partisans—not in changing their vote preferences, but in reducing the likelihood of their casting a ballot. Both effects permit democracy to update on new information.

## The Diversity of Citizens in Democratic Updating

The efficacy of democracy in avoiding bias depends on at least some voters shifting in the face of evidence, either by changing their votes or by deciding to vote or refrain from voting. Democracy could not easily use information to improve government policies if citizens were all equally biased, equally subject to manipulation by frames, or equally ignorant and thus subject to manipulation by special interests. But people differ in such tendencies, permitting some citizens to update, thereby changing electoral and policy results.

First, voters differ generally in how well informed they are.[33] Those who are better informed are more likely to update on the basis of new information than those who are not well informed.[34] Some political scientists have also found that the well informed are less likely to be subject to strong framing effects, presumably because their knowledge allows them to put the frames offered by politicians in a larger context.[35] In experiments citizens who were less trusting of their party did not engage in motivated reasoning.[36]

Citizens also are affected by specific personal experiences that give them special capacity to update better on particular issues than the average citizen. A recent study, for instance, has shown that Republican candidates for president in 2004 and 2008 progressively lost votes among military personnel, a group in which Republicans had previously enjoyed high levels of support.[37] These voters' personal experience and specialized knowledge of the wars in Iraq and Afghanistan allowed them to update their beliefs about the competence of the Bush administration's handling of military affairs. This example shows how citizens with particular knowledge drawn from their daily lives can update despite biases. The finding is consistent with experiments in which it has been discovered that motivated reasoning declines when the cues on a political issue are not themselves partisan.[38]

Such personal experience also has an effect on the power of framing. Citizens are more likely to resist a political frame if their own experience provides a different frame. In general people are more likely to be influenced by the frame if the framer is credible.[39] A frame generated by politicians is likely less credible than one that is constructed from a person's own experience.

In a good society, then, politics continually bumps up against other social spheres that serve as a constraint on bias through the injection of relevant information about politics in a nonpolitical context. Thus, maintaining a vigorous civil and market society where people form views apart from partisan politics helps assure that new information improves collective decision making. The Internet, with its specialized blogs on policy issues, blogs that are often not intensely partisan, is now an important part of this kind of society.

It is said that the best-educated voters tend to be the most ideological and therefore have trouble updating, whereas swing and independent voters are not well informed. But this general tendency does not mean that information will not make a difference. Not all well-informed voters are partisans, and thus information can change these voters' minds. In the 2008 campaign, the greatest shift to Obama came from the best-educated voters.[40] Moreover, as discussed above, information may change the likelihood that even partisans will vote, lowering the turnout of those with ideological positions about which adverse information has been revealed. Finally, as the example of military personnel shows, even citizens who are not generally well informed may shift when information comes out that relates to their personal experience.

Besides differing in the degree of information they possess on various issues, individuals differ in the strength of their biases and thus in their ability to update.[41] Cognitive ability seems to temper confirmation bias, the bias that tends to make individuals look for and consider only information that is consistent with their prior beliefs and worldviews.[42] Individuals have different tendencies to epistemic closure or openness depending on their personality traits.

The differing susceptibility of citizens to bias is consistent with the observation that in elections many independent voters, and even registered party members of weak partisan attachments, are open to persuasion.[43] Some political scientists have noticed an increase in polarization between Republicans and Democrats in the United States,[44] while others have suggested this general tendency has been exaggerated.[45] But in any event an increase in polarization can coexist with the same or even increased democratic updating capacity if independent voters have also increased, because these voters can more easily swing from one side to the other.

The electorate's reaction to the absence of weapons of mass destruction in Iraq shows the power of new information to change minds and results even in the face of partisan bias. It is true that most Republicans continued to support the Iraq War even after information made clear that Iraq lacked the weapons of mass destruction that had served as the principal justification of the war.[46] But updating occurred even among Republicans. Those who were less committed partisans and less motivated by religion changed their minds more than those who were more committed and more religious.[47] As a result of such changed views among both Republican and independent voters,[48] the shape of electoral politics was transformed. Changing views of the war were a major cause of Republican losses in 2006; in fact, President Bush reacted to the losses by dismissing Secretary of Defense Donald Rumsfeld.[49]

This example also shows why dividing people into rigid groupings with different worldviews, as cultural cognition theorists tend to do, does not well describe the dynamics of politics.[50] Because of the diversity of human psychology and the diversity of experience outside of politics, political preconceptions are better described as changing along a spectrum rather than fixed into hermetically sealed worldviews. Moreover, because citizens inhabit many social spheres, they also are subject to influence by people with worldviews that are different from their own.

Both the change in the views of military personnel and the more general shift of the electorate in response to what happened in Iraq illustrate another important limitation on biases and frames. When asked what the most important influence was on politics, British prime minister Harold Macmillan is said to have replied, "Events, dear boy, events." Events temper the power of any comprehensive political worldview or bias. Military people may tend to prefer a hierarchical worldview, which inclines them to vote Republican, but the events in Iraq demonstrated an incompetence on the part of the Republican war leadership that changed the political expression of their worldview.

Politicians may try to reinforce national security frames that favor their party, but real-world events create unanticipated facts that derange these frames. Academic experiments cannot easily replicate either specialized personal knowledge among particular groups of citizens or dramatic world events. As a result, these experiments cannot fully capture the dynamism of democratic updating.

Another salient fact about politics limits the effect of entrenched biases and permits updating: generations are always rising to replace those that are dying. And rising generations are in the process of forming their beliefs rather than fitting new information into past beliefs. Cognitive psychologists confirm that the young are less fixed in their beliefs.[51]

An example of the dynamic of generational updating is occurring in the attitude toward same-sex marriage, an issue upon which opinions would seem to be entrenched by worldviews. Younger people of widely differing ideological views are proving to be more sympathetic to same-sex marriage.[52] The changing assessment of same-sex marriage has occurred as important information has come about homosexuality, notably that it is largely not a choice but an inborn trait,[53] and that freedom for homosexual practices has not given rise to crime or disorder. Thus, even if many individuals are not likely to change their views, society as a whole may update as its composition changes.[54] Generational change is particularly important to constraining status quo bias, because new generations are not as attached to preserving a world they have not constructed.

It might be feared that, whatever its advantages in reducing other kinds of bias, democracy strongly reinforces knowledge falsification because it provides incentives for political actors to suppress opinions that run against majority views. But political majorities are composed of various factions. For instance, Republican and Democratic parties are effectively coalitions. Libertarians and traditional conservatives are often Republicans; traditional supporters of the welfare state and welfare state reformers are Democrats. As a result, party control does not necessarily produce strong support for a comprehensive ideology. Moreover, independents are often uncertain of their views and open to persuasion, particularly when some event calls the majority's views into question. Thus, the impermanent and shifting nature of political majorities tempers long-term falsification of knowledge.

## Constraint of Bias through Political Representatives and Experts

Representative democracy also creates specialized roles that constrain bias. Most obviously representative democracy creates political offices. This creation gives the holders of such offices something to lose—their offices as well as their reputations—if they make decisions based on bias rather than updating on information and consequently make more mistakes. Politicians also have to explain their decisions to people of heterogeneous views. Both the stakes and the audience can decrease biases.

Democratic accountability thus comports with those aspects of cognition that psychologists have shown to improve the updating of information. First, individuals update better and avoid bias if they face possible adverse consequences from their decisions. That proposition accords not only with common sense but also with strands of the cognitive literature.[55] In particular, higher stakes lead people to take time to evaluate the new information.[56] Second, individuals are also likely to perform more rationally if they recognize they will have to justify their decisions to people with a wide range of views, because then they will not "tailor their decisions to the views of the prospective audiences."[57] Partisan gerrymandering currently permits the majority of representatives to run for political office in ideologically homogenous districts and thus avoid both substantive and deliberative accountability. However, this feature is not inherent to democracy, and it can be reformed.

In a democracy where elections are substantially contested, basing a policy on beliefs that turn out to have bad consequences can at least in some cases result in politicians losing their jobs. Even if a representative has a strong worldview, it is in his interest to consider the consequences of policy, because those consequences may affect his electoral prospects and he must explain his vote to people of differing views. Similarly, he has incentives to avoid being blinded by his own frames. Even current majority opinion will be weighed against future consequences, because any experienced politician recognizes that the majority may all too easily forget that it preferred a policy if that policy turns out to have bad consequences.

It might be objected that substantive accountability is absent because voters cannot hold politicians accountable if the voters themselves cannot evaluate the relation between policies and consequences. Not all voters are ignorant of policy, particularly in the long run. So long as they update on consequences, they will provide incentives to politicians to contain their biases. Moreover, though voters may be ignorant of policy details, they are not ignorant of policy effects.[58] For instance, the political science literature suggests that if economic growth is low during the president's term, voters are more likely to vote against him and successors of his party.[59] If local politicians watch rising crime rates without pursuing successful counterstrategies, they are less likely to hold office in the long run.[60] To be sure, because of ignorance, voters are far from infallible in their judgments, but even the limited accuracy of voters makes politicians somewhat accountable for their votes.

It is sometimes suggested that citizens cannot easily update on policy issues because political campaigns lead them to unduly extrapolate from their perceptions about candidates' personal characteristics to perceptions of their competence and ability to correctly evaluate policy consequences.[61] But democracy has a way of tempering this bias as well; campaigns reveal personal frailties in most candidates, rendering character alone a more difficult basis on which to vote.[62]

Collective decision making also generates another distinctive position: the expert. Experts are concentrated sources of information about an area of policy. They have no formal role in our political system, but their views can be extremely important.[63] They help set the agenda by describing the likely results of policy, influencing political representatives, the media and ultimately the people.[64] They wield particular influence on the details of policy that others cannot be bothered to understand. Indeed, representatives and experts have such a large combined effect on policy that some conceptions of democracy understand competition between elites (both representatives and experts) as the essence of modern democracy.[65]

Experts have some internal checks on bias that allow them to deploy their influence in a way that checks bias in society as a whole. First, experts in an area often agree even on matters over which the lay public remains divided. This fact itself provides evidence that information can make a difference despite biases. Economic experts are a case in point.[66] Economists vary widely in their political beliefs, but almost all agree that free trade will create wealth and is a desirable social policy.[67] One reason for their ability to converge even in the face of innate and socially generated bias is that they, like political representatives, have more at stake. They will be subject to professional criticism and possibly ostracism if they ignore the patent facts of their field in the pursuit of some worldview or ideology. They have more influence with their peers if they are perceived to be more objective.[68]

Moreover, the long-term effect of bias on experts is diminished by the same phenomena constraining the long-term influence of bias on citizens. New experts who are less resistant to change are always coming into a field as old ones are slowing down or dying. Indeed, this process may be accelerated by the incentives that new experts have to overturn received wisdom that is faulty. By slaying their elders academics clear a path to renown within their field. The accelerating tempo of life today speeds up this academic cycle no less than the news cycle, rendering each newly minted crop of graduate students a potentially powerful corrective force.

Experts are hardly perfect. Even if bias does not preclude convergence on core issues of their field relevant to current policy analysis, other areas of the field are less clear. Bias thus can certainly prevent experts from updating in the most effective way on some issues.[69] It is also true that there is substantial ideological skewing in many expert fields. Currently professors in the social sciences predominantly lean left.[70] But even with bias, experts make arguments that must meet professional norms, and this structure of discourse helps set boundaries to bias. Moreover, a free society corrects for the institutional bias of experts in a variety of other ways. Think tanks have grown up to provide experts with ideological leanings that are somewhat different from most of those in the academy, and these experts help keep academics honest.[71]

Thus, the incentives of political representatives and experts can sometimes correct for biases in the larger polity. The effect is particularly important for biases that a majority of people naturally hold. Citizens tend to think that foreign trade harms long-term growth and employment, even though the consensus of the expert economists is strongly to the contrary. Despite these sentiments, nations in the West have progressively opened up trade since World War II. Experts have influenced policy makers on the subject. They have kept alive the memory of past mistakes like the Smoot-Hawley Tariff increase of 1930 as an example of a policy that, however popular in the short term, retarded economic growth so much that it proved costly to the reputation of the politicians who adopted it.[72]

Of course, even with majority rule, representation, and the influence of experts, democracy often takes a substantial period to update on new information. But contemporary educational reform shows how updating can begin to happen even in the face of bias.

## Democracy's Capacity to Act in the Face of Bias

In a democracy, political representatives act as the coordinators of expertise,[73] combining the concentrated knowledge of experts with the more dispersed knowledge of their constituents. This combination of the interests of legislators and experts in their jobs and reputation, together with the role played by more open-minded voters, can provide a powerful impetus for reform even in areas where biases have substantial sway.

Currently we are witnessing movements to substantial reform in K-12 education. Although these various reforms are based on evidence and inference from evidence, they do not yet rest on conclusive proof. Yet the recognition of the failures of public education has moved the middle to consider such experiments despite substantial entrenched bias of various kinds.

In 1983 the National Commission on Excellence in Education published a report called *A Nation at Risk*, detailing a decline in the performance of educational institutions.[74] It took a substantial time for this alarm to make educational reform central to the American political agenda. But reports of America's failings, particularly in comparison to other nations, continued.[75] Such reports struck a particular chord because U.S. citizens, whether or not they hold entrenched worldviews, largely share an interest in imparting high educational skills to their children. A better-educated population is perceived as necessary to compete with other nations and to produce the goods and services that will sustain the economy and the future of entitlement programs.[76]

By 2000 these concerns had become so prominent that education featured prominently in the campaign for the presidency.[77] After the election, George W. Bush negotiated with Democrats, and Congress overwhelmingly passed the No Child Left Behind Act.[78] In the bill Congress has encouraged public schools to evaluate the success of their programs based on scientific evidence.[79] In the related Education Sciences Reform Act of 2002,[80] Congress created a federal agency to produce data about and analysis of programs to improve educational outcomes;[81] specifically they called for randomized trials to test educational programs.[82] Thus, despite pressures from citizens of different worldviews and from concentrated interest groups in the form of the teachers unions, Congress established structures to see which kind of reforms will actually deliver better education.

In particular, some of these reforms are already gaining popularity, even though they go against the grain of prevalent biases. A case in point is merit pay, the practice of paying teachers according to a metric of performance rather than by seniority. Merit pay is deeply unpopular with teachers unions.[83] Because unions are run by majority rule, they tend to try to steer maximum benefits to the median member, not the most productive member. Thus, the interest group with the most at stake on the subject can be expected to generate slanted information on the issue. Merit pay also does not comport with an egalitarian worldview; it presumes that some teachers are more talented than others and moves to reward them.

Yet despite these pervasive biases, merit pay is moving to the top of the political agenda of education. Expert studies have been accumulating evidence that merit pay offers the promise of improving educational outputs. These studies have shown a wide variation in student gains depending on differences in teachers' abilities.[84] But teachers' ability is unrelated to education certifications or other fixed inputs.[85] This combination suggests that educational outputs may improve if teachers have incentives to be better or if pay structures attract those who are confident that they will be rewarded.[86] A smaller body of research suggests performance-based pay actually induces higher performance.[87] To be clear, it is hardly yet proven that merit pay will be systematically useful, just that it shows sufficient promise to justify further experimentation.

Reacting to such evidence, in 2006 Congress appropriated $99 million to fund experimental performance-pay programs in states, school districts, and charter schools.[88] President Obama's administration also has been supporting merit pay, despite the fact that teachers unions support the Democratic Party.[89] This stance provides a salient example of a party bucking its own interest group because of concern about the future consequences of policy of interest to voters. At the state level, major performance-pay pilot programs have been developed in Colorado, Texas, Florida, and Minnesota.[90] Public opinion has been shifting so as to welcome these experiments. A very substantial majority of citizens now favor merit pay and believe an educator's pay should either be closely tied or somewhat connected to performance.[91]

Charter schools are an even more important development. A charter school can be broadly defined as a publicly funded school that is operated independently of the state or local school system. There are a variety of mandates and restrictions on their operations depending on the locality, but charter schools generally enjoy substantial independence in setting curricula, choosing and compensating teachers, and establishing a distinctive student culture.[92] Perhaps even more than merit pay, charter schools are thought to have promise. They can offer approaches that may better meet the demands of a particular type of student and are free to make experiments without much regard to the mandates of bureaucratic uniformity.[93] This kind of independence may not only improve the performance of students in the charter schools, but it may also put pressure on public schools to improve by increasing the competition.

Like merit pay, charter schools bump against entrenched biases. Their freedom sits uneasily with a hierarchical mind-set. Their independence and flexibility in dealing with teachers does not comport with the interest of unions in controlling the fundamental relations of employers and their members. Centralized government school systems have been the norm for many years, so introducing charters is also in tension with the bias toward the status quo. Yet charter schools are multiplying. These schools are encouraged by the federal program Race to the Top. They are ongoing in nearly every state,[94] with varying degrees of success.[95]

As with merit pay, the point here is not to show that charter schools are necessarily the last word on education reform, but rather that democratic society can try out plausible reforms that fly in the face of biases against them. Moreover, charter reforms are themselves structurally consistent with a politics of learning. Even those scholars who find that most charter schools perform no better than their public school counterparts also find that some charter schools produce substantially better results.[96] Because the approaches of charter schools by their very nature differ from one another, they create a trial-and-error approach to reform. Results can be measured and other charters (and to some extent traditional public schools) will copy them. Thus, charter schools provide a structure of reform that is consonant with modern information technology. Other things being equal, government should choose reforms through which measurements of differential success can lead to a policy dynamic of good reforms driving out bad.

## Information Technology and Bias

Charter schools thus illustrate the relation between government structure and bias. The more government can be restructured to provide usable information about policy effects, the less bias there is likely to be. It is rational to vote on the basis of party labels if it is difficult to find out more information about the consequences of the actual programs that a politician advocates. But the easier it is to discover the actual consequences of policy, the less need there is to rely on political labels as proxies for what the results of policies are likely to be. Thus, information-eliciting rules, together with empiricism, prediction markets, and dispersed media, are likely to have a feedback effect in making voters less reliant on information shortcuts, such as party labels. Indeed, it turns out that party identification has been decreasing among the electorate.[97]

More generally, new information technology can create a virtuous circle, prompting demands that it be used for transparency and better information production about politics. Examples already discussed include the requirement that bills be made available on the Internet days before they are voted upon or signed and demands that campaign contributions be disclosed immediately. Such changes then make it easier for the public to mobilize on behalf of further reforms. This kind of virtuous circle is nothing new in the history of information and politics. The printing press, for instance, made it easier for the public to stand up to the privileges of autocratic rulers. The feedback effect of information technology on politics has worked slowly in the past. But as the example of education reform suggests, it may well be accelerating today, despite the entrenched interests and biases that impede reforms.

Nevertheless, bias, whether it comes from entrenched worldviews or from the power of interest groups, is unlikely to be dispelled simply by more information. An agenda for accelerating democracy must include political reforms for reducing bias more directly so that the new information technologies can improve collective decision making as much as possible. It is to that agenda we now turn.

# CHAPTER NINE

# De-biasing Democracy

BIAS REMAINS A SUBSTANTIAL IMPEDIMENT to improving policy on the basis of new information. Thus, reforms that constrain bias are an important element of accelerating democracy. Some of these reforms would build on the existing mechanisms of democracy like majority rule for electing candidates and responsive representation that already constrain bias. Other reforms would deploy new information technology to further constrain bias.

## Reinforcing Majority Rule and Representation

### Gerrymandering Reform

Majority rule promotes updating on information in elections, because less partisan and more open-minded voters cast the decisive ballots to choose candidates for office. But partisan gerrymandering can reduce the influence of swing voters. Gerrymandering can also entrench political representatives in office, making it less necessary for them to explain their positions to voters with diverse views, a process that can reduce bias among legislators. Empirical evidence supports the need for such reform. Representatives in districts where swing voters are more likely to be decisive vote in a more independent and less partisan manner.[1]

Under the laws of most states, state legislatures draw up districts both for themselves and for members of the House of Representatives. State legislatures engage in two kinds of partisan gerrymandering that inhibit the influence of swing and independent voters. In one kind a partisan majority in the legislature attempts to elect as many members of its party as it can. It thus packs voters of the other party into a few districts, minimizing the opposing party's representation. In other districts the majority party places just enough of its partisans to gain victory, thus maximizing the number of seats with the minimum of support.

In the second kind of gerrymandering, partisans of both parties create a kind of duopoly to maintain the status quo. Seats are created with enough Republican and Democratic partisans to assure the election of members of each party in districts dominated by their respective partisans. This system often reflects the power of incumbents, because it provides them the maximum protection against electoral vicissitudes

The first kind of gerrymandering is, to some extent, self-limiting. Large electoral swings may endanger and defeat members who have been placed in districts with a partisan majority that is just sufficient to assure victory. In contrast, the duopoly strategy can create large partisan majorities for both parties and makes members relatively impervious to electoral swings.

Technological acceleration has made the problem of partisan gerrymandering more acute. New computer technology allows politicians to draw districts as they wish, permitting them to protect incumbents and their partisan advantage.[2] Thus, at the very time technology is providing a greater capacity for democratic updating, it is also providing the means for creating more opaque filters of bias through which this information is screened.

From the perspective of consequentialist democracy, the real evil of gerrymandering is that it reduces the ability to update on new information by limiting the responsiveness of the electorate to new information.[3] Partisan representation should change as people's assessments of party policies change.[4] Responsiveness is best advanced by focusing on the degree to which the change in the number of seats in the state reflects the change in the number of votes the party receives.[5] Responsiveness is directly related to the competitiveness of districts.[6] If districts are packed with a large majority of partisans to assure election of a particular party's candidate, they will not be very responsive.[7] Gerrymandering reform increases responsiveness by making it more likely that districts will shift from one party to another as voters update on information and shift as well.

Thus, gerrymandering reform facilitates democratic updating by constraining bias. States are showing an increasing interest in making such reforms. For example, for a long time Iowa has redistricted through a commission, a practice that has yielded more responsive districts.[8] More generally, in the eleven states that used commissions to design congressional districts in 2002, more competitive elections resulted.[9]

The attempt to cabin gerrymandering has recently moved to the nation's most populous state. In 2008, by citizen initiative, California established a citizen commission to draw district boundaries for state legislatures.[10]

In 2010 voters extended the scope of the previous initiative to mandate that the commission also set congressional boundaries.[11] The rationale of its supporters focused largely on the benefits flowing from the reduction of partisanship.[12] A likely by-product will be to increase responsiveness, as the boundaries selected to be redrawn have been heavily criticized as entrenching conservative Republicans and liberal Democrats.[13]

In fact, empirical studies of the effects of gerrymandering reform in California over the past forty years show that redistricting, when controlled by commissions, not the legislature, resulted in less partisan bias and that these positive effects of gerrymandering reform disappeared when control was given back to the legislature.[14] So Democrats "tend to take more conservative stances on environmental and labor issues when they represent panel-drawn districts, and when districts are legislatively drawn they take significantly stronger pro-environmental (or pro-labor) stances."[15] California Republicans behave the same way, voting in higher numbers for traditionally conservative issues when their districts are legislatively drawn and acting more moderate when a panel draws the lines. Proposed drafts of the new jurisdictional maps for California have met with some acclaim, as the "unraveling" of gerrymandered districts threatens incumbents, "enticing newcomers ready to unleash their bottled-up ambitions" in the 2012 elections.[16]

In 2010 the citizens of Florida attempted to restrict gerrymandering by constraining legislative discretion rather than empowering an independent commission. Their constitutional amendments required legislators to draw the boundaries of districts "as far as possible in a compact and contiguous manner."[17] Focusing on compactness and contiguity can be a proxy for responsiveness, because these features conflict with objectives of preserving incumbents or entrenching partisan advantage, objectives that undermine responsiveness. Thus California and Florida offer substantially different routes to gerrymandering reform, with California relying on an independent commission and Florida still relying on the legislature, albeit with judicially enforceable constraints. Federalism permits experiments to see which approaches work best.

Gerrymanders are not the only reason why districts become polarized.[18] For instance, citizens may be increasingly sorting themselves into homogeneous residential districts.[19] But confining gerrymandering aimed at a reduction of competiveness should reduce one important cause of district homogeneity.

One possible concern is that by increasing responsiveness, gerrymandering reform may deprive people of representatives in the legislature who have distinctive, even if extreme, views.[20] Dissenters—politicians who are outside the mainstream—can be valuable for democratic updating. Sometimes the consensus is wrong and deliberations are likely to improve when those in the middle of the legislature must justify themselves to colleagues who are more extreme. But even policies that increase responsiveness will not eliminate all political representatives with extreme views. Some people will appeal to extremes and thereby get a large voter turnout of their base. Others will be elected because of their charismatic personalities, whatever their views. Still others will continue to be protected by incumbency. At the federal level moderate voters in Idaho will continue to differ sharply from moderate voters in Vermont. Even within a state, voters in localities so distant that they cannot be plausibly integrated into a single district often have sharply different viewpoints.

### Reform of Primary Voting

Another way of making politics more responsive is to create different rules for election primaries that will help elect more moderate legislators.[21] Currently many states operate closed primaries, which limit their participation to party members. This structure limits the influence of independent and moderate voters. Moreover, because primaries elicit a lower turnout, party activists and more extreme voters may exert disproportionate influence. The prospect of such a primary thus tends to increase bias, because it forces representatives to make decisions with a view to presenting them to a very ideologically homogeneous primary electorate. The general election then often becomes a choice between two polarized candidates, neither of whom appeals to the center.[22]

Some states do permit open primaries, in which voters of any affiliation may vote for the slate of any party, thereby allowing independent voters and voters from smaller parties to shape the results.[23] The influx of such voters should have a moderating influence.

An even more radical step are so-called top two primaries, in which all voters participate in a single primary. Afterward, the actual election takes place between the top two finishers.[24] Such a primary gives power to independents to participate in selecting the top two finishers. This influence in the primary may well make the top two finishers more moderate even when, as is generally the case, they consist of one Republican and one Democrat. But the effect of a top two primary may be even more dramatic in a district so skewed to one party that both top finishers are of the same party. In that case, independents and members of the opposite party can have substantial, perhaps decisive, influence in the general election, because the candidates from the same party cannot ignore such voters as they could in a partisan primary whose result would then effectively determine the winner in the general election.

As with the varieties of gerrymandering reform, the optimal primary reforms are not yet entirely clear and may potentially vary from state to state. Federalism permits experimentation here too. California, for instance, voted in 2010 to shift to a top two primary system.

The Supreme Court should not put any impediment to such primaries. It is true that in 2000 the Court held unconstitutional a so-called blanket primary, in which all candidates ran in one election and the top candidate from each party was chosen.[25] Since then, however, the Court has not extended these principles to invalidate any other kind of primary and has, in fact, upheld rules for a top two contest in the state of Washington.[26] As a part of jurisprudence of social discovery, the Court should allow states to experiment with different versions of these rules.

## Earmarks

Earmarks are the practice by which individual members of Congress target appropriations for their own district outside of any competitive process or other neutral criteria.[27] The usual complaint is that earmarks—a kind of pork barrel politics—add to the budget deficit[28] and green-light inefficient projects like the infamous Bridge to Nowhere in which the costs far exceed the benefits to the nation as a whole.[29] Recently the House of Representatives instituted a ban on earmarks, and the Senate reluctantly followed suit.[30]

But a ban on earmarks also increases responsiveness to policy issues and curbs bias. Earmarks interfere with responsiveness in two ways. Since the ability to gain earmarks increases with seniority,[31] earmarking gives voters reason to support the incumbent, whatever the incumbent's policy positions.[32] Earmarking also creates an axis of electoral competition that is irrelevant to democratic updating on policy. The practice raises the political salience of the question of who can bring home the bacon rather than who is pursuing the best policies more generally.[33] Candidates can gain great advantage from succeeding on this axis of politics, because voters more easily associate a specific, visible local project with a candidate than with a complex social policy.[34]

The de-biasing rationale against earmarks strengthens the case for their abolition, because this rationale surmounts many criticisms of the ban. The earmark ban has been dismissed on the grounds that the amount of money saved is trivial compared to government spending as a whole.[35] But even if the amount of money affected is small, the benefit for democratic updating can be large because of the substantial aid earmarks provide to incumbents. Some suggest that the amount of money saved will be even smaller than it seems, because communities and members of Congress will lobby executive agencies for funds.[36] But lobbying behind the scenes is harder to attribute to a member of Congress, even if he or she is the lobbyist, and thus does not substantially detract from the democracy-reinforcing nature of eliminating earmarks.[37]

Congress should make the current moratorium permanent. State legislatures should do the same for earmarks that use state funds to benefit particular local districts. The power to create local projects can also benefit incumbents and make it more difficult for citizens to update on information.

## Term Limits

Concern about bias also provides new rationales for term limits for legislators and new impetus for state experimentation in this area.[38] The first of these rationales builds on more conventional concerns about the advantages of incumbency that have long motivated the public's interest in term limits. A long-term incumbent has generally enjoyed political advertising and other publicity over the years to create name recognition, which gives him or her an advantage over challengers. A long-term incumbent can be analogized to a well-established product with substantial barriers to entry for any newcomers.[39] Indeed, the advantage is stronger in the realm of politics than in private markets. In private markets, citizens have strong incentives to search for new products, because their buying decision can guarantee that they reap the advantages of discovering something better. In politics, however, voters have fewer incentives to look for something new, because their own voting decision does not guarantee that they will gain a new representative. Moreover, even with the elimination of earmarks, incumbents enjoy advantages, such as seniority on committees, that may deliver benefits to their constituents.[40]

Name recognition and the prospect of constituent benefits tend to slant elections in favor of incumbents, even in elections where incumbents are thought to be unpopular. In the 2010 midterm election, touted as a major coup against established politicians, only 53 of the 435 incumbents lost their House seats.[41] The biases created by incumbency make it less likely that the election will be decided based on democratic updating of policy.

Additionally, the imposition of term limits may increase the cost of bargaining for special interests, thus reducing their influence on the government. Without the guarantee that an incumbent will stick around into the indefinite future, special interests would be less confident that their favors will endure, thus decreasing their incentives to inject false and self-serving information into the political process to reap the rewards of a long-term deal.[42] Term limits set a boundary to the duration of the benefits of name recognition and the prerequisites of office, making elections more likely to be decided on policy positions. Thus, the traditional justifications for term limits also provide reasons to support them to lessen bias in elections.

But the status quo bias discussed in the last chapter offers a wholly new rationale for term limits. Term limits would tend to reduce the number of long-serving political actors, thus reducing the age of those who must deliberate on the effect of new evidence on social policy.[43] In so doing, term limits would create a younger set of decision makers, who are less likely to suffer from status quo bias.

Age among members of Congress has been rising relatively steadily since 1951.[44] The average age now stands at sixty for Senators and fifty-seven for members of the House.[45] These raw statistics understate the power of the elderly in Congress. Because of the seniority system, chairmen of committees are very disproportionately drawn from the ranks of the oldest. The median age for chairmen of standing Senate committees is sixty-nine. The chairman of the House Science Committee, the committee one might think is most concerned with the future, is Ralph Hall, who at the time of this writing is eighty-nine years old.

If term limits were imposed, the median age would fall, because the median age of newly elected legislatures is much lower than that of the legislature as a whole; for instance, the average age of senators newly elected in 2010 was fifty.[46] Even more important, term limits would retire that group of representatives and senators who, elected term after term, become superannuated and are most likely to be affected by status quo bias. Of the twenty-one senators who are currently over seventy, all but one has served for more than two terms. We have less data on the ages of state legislators, but we do know that approximately a quarter of them are over sixty-five years old.[47] A younger legislature may be more important than a younger electorate in reducing the effect of entrenched bias on decision making, because, as we have seen, in a representative democracy legislators may be in a better position to anticipate change than individual citizens.

Given that technological acceleration will inject wholly new kinds of issues into the body politic, reducing the influence of status quo bias and general ignorance of contemporary technology is more important than ever. An aging legislator may well have difficulty putting aside the prejudices of a lifetime to think afresh about a policy issue that has only just come to fore.

Unfortunately, a closely divided Supreme Court struck down a state law that imposed term limits on federal legislators on the grounds that it unconstitutionally added to the few qualifications the Constitution expressly imposes on federal legislators.[48] The idea that federal legislators will propose a constitutional amendment for term limits is as likely as turkeys voting for Thanksgiving. Thus, the only plausible route for a constitutional fix would be for state legislators to call for a constitutional amendment. Fortunately, the Supreme Court's decision does not prevent states from setting term limits for their own officials. A movement for a federal convention on this subject could gain momentum if state experiments with term limits demonstrate their benefits. Thus, state experimentation and the measurement of results may once again be the key to crafting the most effective term-limit reform.

The case for term limits may be more pressing for the Supreme Court itself than for legislators.[49] Even today, after four appointments in the last six years, Supreme Court justices already have a median age of sixty-three with four members in their mid-seventies.[50] Moreover, with better medical technology, the justices are likely to retain their seats longer. If one is a pragmatist about law, policy considerations will be essential to making constitutional and statutory decisions,[51] and status quo bias will be an obstacle to updating on the facts relevant to policy. Even a formalist justice still must apply laws to facts, and factual updating will proceed better with less bias, because the justices will be more responsive to changes in the facts.

Regardless of judicial philosophy, it has been well observed that modern Supreme Court justices have become legal celebrities who do not deliberate together as much as follow a personal jurisprudence that they formulated early in their tenure and with which they continue to identify.[52] Term limits would help assure that their jurisprudential stance would be adopted closer to the time of the judicial rulings in which it would be applied.

## Education Reform

Changes in education might also make democracy more effective in constraining bias. This connection between a particular kind of democracy—one where people are less biased—and a particular kind of education is hardly surprising; an educated electorate has long been thought necessary for an effective democracy.[53] What is new is the notion that specific kinds of education may be useful to de-biasing citizens, thereby making them more dispassionate consumers of information.

For instance, an education in the content of biases may help people become less biased.[54] There is also evidence that teaching students to develop the habit of looking at counterfactuals—what would occur if a policy or event did not happen—is useful in coping with biases, because it encourages consideration of alternatives, making for a more fluid mind-set.[55] An education in statistics today can empower a wider number of citizens to evaluate empirical findings and become comfortable with prediction markets, which are themselves a tool of de-biasing.[56]

Of course, the time and resources for education are not infinite. Education for the containment of bias comes at the expense of some other sort of education. But given the high stakes of social decision making, psychologist Steven Pinker, for instance, has suggested that schools should focus more on providing tools to combat the biases that make collective decision making less accurate.[57] Moreover, technological acceleration may make some academic subjects less necessary. Machine translation, for example, will enable people to understand foreign languages without learning them.

Some critics have been skeptical of the benefits of more education to democracy. It is true that advances in education since the 1900s have made the electorate only marginally better informed about political facts.[58] But the focus here is to make the electorate less biased in the use of information rather than better factually informed. Political scientists have suggested that improved education has rendered the electorate more skilled about making the connections between different political events, like that between deficits and inflation.[59] Dispelling bias also improves one's perception of connections—in this case, the connections between one's own biases and evaluation of policies.

## Prediction Markets—Again

In addition to injecting more information into collective decision making, prediction markets can help with bias. Prediction markets reduce the ideological bias among experts by subjecting their predictions to a market test. They also reduce ideological bias among the citizenry by forcing citizens to consider the possible alternative results of policies. Finally, they reduce knowledge falsification, because they motivate people to make predictions against majority opinion.

First, conditional markets could be useful in combating bias among the general citizenry. As we saw in chapter 8, cognitive scientists have suggested that forcing individuals to consider the alternative is one way of combating bias. Conditional prediction markets by their very nature can force the consideration of alternatives. For example, a conditional market on a capital gains tax cut forces people to consider the alternative economic situations in which a capital gains tax is cut and in which it is not, thus forcing both proponents and opponents of capital gains tax cuts to confront the preferred policy world of the other and observe the predicted results. A market considering both conditions also prompts consideration in a relatively nonpartisan context. The absence of partisan cues reduces the bias of motivated reasoning.[60]

Second, prediction markets create a market check against a different kind of bias—the bias of experts—because a market can be made on their predictions. This market would then allow citizens to bet against expert predictions if they chose. Prediction markets thus could institutionalize challenges to expertise on a wide range of policy issues like the effects of tax cuts or health care policy. These markets can provide a check on experts, making it more difficult for the peer review of other experts to insulate claims from contestation.[61] This check in turn may make experts more wary of indulging in ideological bias.

Third, prediction markets provide incentives for people to inject their opinion into the body politic, even when it is contrary to that of the majority. Society's knowledge runs the risk of falsification if a strong majority can suppress dissenters from expressing their view of the adverse consequences of the majority's policy. But markets provide a financial incentive to run apart from the herd and get off the bandwagon. If the markets are anonymous, bettors against received wisdom need suffer no social opprobrium at all.

Bias has long been seen as a threat to accurate government decision making; the adage that no man should be the judge in his own case is an early reflection of this concern. Over time society has acted to prevent personal bias, such as a direct financial interest, from distorting the rulings of judges, the decisions of administrators, and the votes of legislators. Greater transparency in government has decreased bias that favors special interests. But now society has accumulated knowledge about more subtle yet pervasive biases—from biased assimilation, to knowledge falsification, to status quo bias. We should use this developing social knowledge to create better mechanisms of constraint against bias and thereby make new information about substantive policy more effective for democratic updating. An age of technological acceleration can less afford bias than previous ages, because its speed of social change may make mistakes less easily correctable.

# The Past and Future of Information Politics

TODAY'S NEED TO MATCH SOCIAL GOVERNANCE to technological change is but the latest chapter in a long story of political adaptations to material innovation. Improvement in social governance has tracked the improvement of the political information sphere—the sum of political institutions that facilitate the creation, distribution, and use of social knowledge. Improvements in the information sphere have in turn depended on technological innovation. In the information age, information can now be readily seen as a driving force of history, one that is probably more important than race, ethnicity, class, or religion.

The history of the symbiosis of information, technology, and governance cannot be written here. Nonetheless, a few vignettes from ancient Athens, Britain on the cusp of the industrial age, and our country's own founding can display the arc of change, the manner in which successful societies exploit the technologies of the time to construct a more productive information sphere to better assess and predict social policy. Despite the differences among these societies in time and place, their successes depended on the same kinds of reforms necessary for our own society, such as creating a more powerful compound of expert and dispersed information, distributing social knowledge more efficiently, and giving political actors better incentives to use it. These sketches thus confirm that our need for better mechanisms of social assessment is not unique to our age. They also underscore that even if we adopt reforms today, new information mechanisms will be needed tomorrow.

## From Technological Invention to Innovations in Social Governance

Throughout the history of the world individuals have invented new devices for human benefit. These inventions in turn alter the relations of people to one another, creating opportunities and needs for new forms of political structure. Thus, accumulating acts of individual genius often end up transforming collective governance.

In part, technological innovation itself generates the need for new, more information-rich forms of social governance. As Brian Arthur observes, "Every technology contains the seed of a problem, often several. . . . The use of carbon-based fuel technologies has brought global warming. The use of atomic power, an environmentally clean source of power, has brought the problem of disposal of atomic waste. The use of air transport has brought the potential of rapid worldwide spread of infection."[1] It was always thus, even from the beginning of recorded history. The technology that created agricultural surpluses made cities possible, generating new problems, like the disposal of human waste, that called for collective solutions. By accumulating wealth within a compact space, cities also attracted marauders, requiring better policies for defense.

But besides such specific problems, technological innovation generates more general difficulties for social governance, because such innovation renders the social environment more and more distant from that in which we were adapted to live. In the evolutionary era, humans inhabited small communities where members were related by sexual bonding or by blood to many other members. Technological innovation, however, has progressively increased the gains from trade and specialization that come from living in larger and larger groups. As the polity moves from the tribe, to the city-state, to the nation-state, and perhaps in the future to a more international structure of governance, society can rely less on the fellow feeling of extended kin and ethnic groups to reach agreement and maintain stability. With specialization and innovation, society also becomes more complex and the relation of its constituent parts often becomes difficult to comprehend. As Ian Morris notes in his history of the last millennia, "The price of growing complexity was growing fragility. This was, and remains, a central piece of the paradox of social development."[2]

But if technology creates problems for social governance, it also creates opportunities to improve the production of public goods though better mechanisms of information. One of those opportunities is better coordination for defending society against enemies outside and criminals within. Since technology may improve the coordination and power of attack as well, defense against outsiders is one opportunity that societies must exploit in a world of competing centers of power. The new social structures made possible by technology can also increase the effectiveness of other public goods besides defense. For instance, public education offers a source of prosperity by generating information spillovers that increase economic growth. But to gain these advantages society must make the right collective decisions about such public goods.

The new problems and opportunities for social governance must be addressed by pooling information. An individual working largely on his own can change the world by material invention. But any one individual may find it more difficult to understand the primary, secondary, and tertiary effects of public policies as they ripple throughout the social world. Moreover, inhabitants within a polity disagree about what these effects are. These disagreements often reflect not only differences in factual knowledge but also differences driven by ideology and parochial interest. Thus, while technological advance is often spurred by solitary genius, social governance is best advanced collectively by creating mechanisms for capturing more dispersed knowledge within society.

## Athens: The Beginning of the Modern Information Sphere

With good reason, the democracy of ancient Greece in general and Athens in particular has long been understood as the beginning of Western progress in social governance. Greek democracy itself was made possible by technology. Improved methods of agriculture like the iron plowshare created a surplus to support cities.[3] Cities in the Aegean could prosper as well through specialization, because improved methods of shipping permitted trading for goods that were made more efficiently elsewhere.[4] With such commercial surplus, citizens in Greek city-states had more specialized occupations and thus became repositories of knowledge that can be considered expert for the time.

As Steven Johnson notes, cities "cultivate specialized skills and interests, and they create a liquid network where information can leak out of these subcultures."[5] But even if the technology of cities creates information spillovers, political structures must be built to take advantage of them. According to Josiah Ober's magisterial study of democracy in classical Athens, Athenian democracy was more successful than other city-states at producing public goods, from defense to monuments, because of the way its social structures produced social knowledge.[6] Of course, over the long duration of Athenian democracy there were setbacks, notably at the end of the Peloponnesian War. But Ober examines Athens over a period of many centuries and finds that it generally outperformed other city-states.[7]

Most famously Athens became a direct democracy where primary decision making was made in an assembly of approximately eight thousand male citizens. The assembly itself gathered the dispersed knowledge of citizens.[8] But such a group is too large for day-to-day administration. Moreover, its only intermittent attention to public affairs inhibits the setting of agendas and the gathering of expert information. Ober shows that Athenian democracy crucially relied on the Council of 500, a complex social structure for information networking that performed these functions.[9]

The members of the council were picked by lot from particular areas of the city-state. They then lived in the city proper for a year and were assigned to fifty-member teams or committees within the council. These teams created a network for exchanging information across "regions, kinship groups, and occupational groups."[10] Through this network, members could gain expertise solving the problems of the polity, creating an early example of the compound of dispersed and more expert knowledge that is so necessary for setting democracy's agenda. The members also had incentives to put this compound to use, because a well-regarded member of the council would improve his "social network" on his return to public life. Thus, long before the rise of prediction markets and dispersed media helped to combine expert and dispersed methods, societies tried to accomplish similar objectives with less advanced mechanisms.

Ober shows that other features of Athenian democracy were designed to facilitate social knowledge as well. For instance, laws were written in easily understood language and prominently posted in public forums.[11] Athens thus made use of additional information mechanisms that were consistent with the technology of its time, which gave its members access to social knowledge.

## Britain and the Industrial Age: Representative Democracy, Civil Society, and the Information Sphere

As technological innovations in such matters as agriculture, cloth production, and energy capture made societies even wealthier and more mobile, nation-states began to become more salient forms of organization than city-states. Such states provided even greater opportunities for specialization and gains from trade. But technological advances and the structure of the nation-state also offered the possibility of innovations in constructing a more productive information sphere in politics.

First, the larger size of the nation-state naturally led to a theory of representative democracy and the use of elected legislatures rather than popular assemblies for democratic decision making. In a nation-state it is impossible to gather citizens together in one place. But representative democracy has other advantages in information production and deployment over the direct democracy practiced by ancient city-states. The information available to direct democracy is limited by the time and place of voting, whereas a parliament can tap into a more continuous stream. Moreover, legislatures can acquire cumulative knowledge of substance, or at least awareness of where to acquire substantive knowledge.

Parliamentary forms of government also have their own characteristic deficiencies from an information perspective. Most notably, the legislature may become insular and remote. But at the same time as the potential for parliamentary forms of government increased, the rise of the printing press, itself a product of progress in mechanical technology, made possible more sophisticated and reticulated opinion to which legislators concerned with reelection must heed.[12] The press also permitted deliberation to stretch over territory and time.

Just as Athens was the society that constructed the best information sphere in the city-states of its time, so England, and then Great Britain as a whole, was the leader in constructing an information sphere for politics in its time. As Joel Mokyr notes, the Parliament in Britain was uniquely powerful.[13] It also, for its time, represented the greatest cross-section of interests of any institution in Europe. It thus brought together reports about the world from every corner of the kingdom, offering more varied information than that provided by systems dominated by monarchies and their courts in most nations of Europe. Economic historians have argued that the result was better laws about property and industry that were conducive to economic growth and indeed the forging of the industrial revolution.[14]

But just as a more powerful and representative parliament injected more information into political decision making, that institution was surrounded by a more vigorous public exchange about policy. While Britain had nothing like the freedom of speech that exists in the United States today, it was substantially freer than elsewhere. The ability to hold and express unorthodox opinions on public affairs was noted by foreign observers like Voltaire.[15]

Britain also generated more sources of expertise to influence policy. Adam Smith, for instance, was appointed as commissioner of customs despite his criticisms of government policy. The work of such experts, both publicly and privately employed, was important in pushing wealth-producing policies like free trade and resisting the vested interests that often try to obstruct technological progress.[16]

This overall structure was rewarded in several ways. First, the information-rich polity helped make Britain the cradle of the industrial revolution.[17] Second, it helped Britain defend itself against more populous nations and become a world power. Thus, the history of Britain provides an example of how government action can increase the arc of progress for social knowledge. The government permitted more criticism of social policy. It provided support for relatively independent experts. It created as its principal instrument of decision making a structure that represented a greater cross-section of people and made them accountable for decisions by holding elections.

All of these features of the British system were imperfect. Until the nineteenth century the British Parliament contained rotten boroughs, constituencies with very small electorates that represented the interests of a particular patron. As a result, they were not likely to update on information. But these rotten boroughs were slowly reformed to produce a more responsive electorate. From the perspective of creating a better information sphere, that process of reform is not different in kind from our movement to reform gerrymandering. The larger lesson is that even imperfect innovation in the information sphere can deliver benefits despite its defects and that these benefits can grow with further reforms. It provides a useful lesson to remember as our society experiments with innovations such as prediction markets that may initially have shortcomings.

## The Founding: The Information Sphere and the Creation of Political Structure

The founding of the American Republic can also be understood as an essential step in the progress of the information sphere. For the first time, the technology of the day permitted a large polity as a whole to participate in consciously choosing its fundamental law: the Constitution. The decisions about structuring this deliberation reflected a concern with combining expertise with dispersed knowledge. In early America communication and transportation costs were falling sufficiently that even a large polity could create the kind of "liquid network of information" that city-states once enjoyed. The speed of transportation was increasing, and thus the time it took for news to travel was decreasing. Newspapers were increasing in circulation and scope, permitting citizens to unite for common purposes. As Alexis de Tocqueville wrote, such organization for the public good "cannot be done habitually and conveniently without the help of a newspaper. Only a newspaper can put the same thought at the same time before a thousand readers."[18]

America's founders took advantage of these changes to promote an intensive debate about consequences though their structure for ratification. First, the U.S. Constitution was a two-step process. Delegates were initially selected to debate the proposals for a new constitution at a national convention. The members of this small group at Philadelphia were or became intimately familiar with one another, much like the Athenian Council of 500. Working without publicity, the Philadelphia convention established the agenda by coming up with a set of constitutional proposals.

But the Constitution was then ratified in state conventions with an unprecedented diversity of citizens from all walks of life. As Boston's *American Herald* wrote at the time of the Massachusetts ratifying convention: "The body now convened is perhaps one of the compleatest representations of the interests and sentiments of their constituents, that ever was assembled. No liberal or mechanic profession, no denomination in religion or party in politics was not present."[19] In each state the ratification debate thrived not only within the conventions but outside as well. Pauline Maier notes that the debate occurred in "newspapers, taverns, coffeehouses, and over dinner tables as well as in the Confederation Congress, state legislatures, and state ratifying conventions."[20] Specifically in Pennsylvania, the state legislature ordered that the Constitution be read publicly to large crowds of citizens and that it be copied in English and German and distributed throughout the state.[21] Such subsidies for knowledge reflected the idea that the citizens of a republic should understand matters pertaining to the public welfare.[22]

Throughout the country, sophisticated students of government discussed the likely consequences of the new design. Crucially, the laws of the day permitted the analyses of the Constitution, both positive and negative, to circulate without fear of government intervention. In fact, the state governments did not apply to this debate the restrictions that the British government had placed on the press, such as criminal libel, government licenses, and onerous taxation.[23]

*The Federalist Papers* are now the best-known essays, but scores of others, both critical and supportive of the Constitution, were published in important newspapers, frequently republished in others, and sometimes collected into freestanding pamphlets or books. These essays canvassed the history of Western nations to find support for their claims about the regularities of human behavior. As befits the first deliberation by a nation over the structure of its political system, the debate reflected, in the words of Alexander Hamilton, a new and improved "science of politics."[24] This science posited that political structures could create incentives to encourage or discourage predictable and desirable political results.[25]

Thus, at the center of the debate about our founding institutions, we see the beginning of an empirical, positive evaluation of political consequences—a kind of inquiry that better technology allows us to update today. While the Constitution is rightly celebrated as a document that generates a better politics though an intricate set of checks and balances, those careful checks and delicate balances were themselves the products of the greatest concentration of dispersed social knowledge for constructing political institutions that the world had ever seen.

Like the information spheres in Athens and Britain, the information sphere surrounding America's founding paid dividends. Indeed, as a result of the debate over ratification, a consensus developed in favor of the Bill of Rights, the portion of the Constitution perhaps most admired around the world. Prominent among those rights were commitments to free speech and free assembly, permitting citizens to construct better information feedback and assessment of policies.

The first party clash in America also revolved in part over how best to build knowledge in politics. As the historian Gordon Wood has detailed, in Federalist thought political understanding reflected an "intellectual unity" that emerged from an "organic unity."[26] As a result, democracy's function was to select office holders from among its elite members and then leave them to apply their knowledge to run the government. The infamous Libel and Sedition Acts penalizing criticism of officials reflected a top-down model of social knowledge.

In contrast, Jefferson's Democratic-Republicans wanted to produce political knowledge through diversity rather than unity. Democracy should rest on an "'aggregation of individual sentiments,' the combined product of multitudes of minds reflecting independently, communicating their ideas in a different way, causing opinions to collide and blend with one another, to refine and correct each other, leading toward the 'ultimate triumph of truth.'"[27] The contemporary discussion over whether it is better to have a culture of political information dominated by the mainstream media or one generated by the blogosphere echoes the earliest political debates of the young republic.

More generally, ancient Athens, Britain on the cusp of industrialization, and the founding of America all shared similar characteristics, as they maximized the opportunities for the production of social information given the technology of their time. They all attempted to mix expert and lay knowledge in a fruitful compound, tap into dispersed sources of information, and provide political actors with incentives to pay attention to the information produced. The information sphere gradually became more productive, because improvements in technology permitted information to be better categorized, sifted, analyzed, and distributed to assess social policy.

## Toward a Multiverse Politics

As this history suggests, the need to make politics more information-rich is an enterprise without any termination date. Of course, new mechanisms for social information cannot be foretold with precision. As with previous transitions, new forms of social governance will take advantage of supply for new technological tools and be driven by a demand for better governance to address technological change. Nevertheless, given the trajectory of what has come before, it is possible to provide a sketch of the future contours of a successful information politics.

The demand side of the equation will likely reflect even greater rates of technological acceleration. Given the prospect for continued exponential technological acceleration, some computer scientists and theorists posit a technological "singularity."[28] In physics a singularity represents a rip in the physical fabric of the universe, such as that caused by a black hole, beyond which it is difficult for outside observers to see. By analogy, a technological singularity creates a rip in the fabric of civilization beyond which it is difficult for observers to comprehend the tenor of human life. The most important marker of this kind of singularity is thought to be computational power equal to human intelligence. Once people design such a computer, its computational power can design further computers of even superhuman intelligence. Because these computers, unlike humans, can work ceaselessly and share information seamlessly, technological change will then be unimaginably rapid. Some theorists put the date for the singularity any time between 2030 and 2100.[29]

Although this idea has been discussed by technological theorists for two decades, it is now becoming more widely accepted as a possibility worthy of study. Google and NASA are funding the Singularity University that focuses in part on preparing for this event.[30] It is not necessary, however, to credit the likelihood of a technological singularity in the near future to believe that the seriousness with which it is taken presages fast change in the next decades.

While the rapid-pace change makes it hard to predict specific technological change, the very speed foretells its capacity to disrupt society. The key to improving social governance to handle these challenges will be to accelerate current trends in the projection of alternate worlds in order to help assess the future policy consequences for the actual world. Alternate worlds can help us better assess both the past and future political results. They also help reduce bias in processing that information by forcing consideration of alternatives. A useful name for this effort is *multiverse politics*, because it tries to evaluate the policy paths in many states of the world to better understand which one we should choose in ours.

Multiverse politics can most obviously develop through the multiplication of prediction markets. Prediction markets create alternate worlds by assessing the results of alternate policies before they are implemented. As they provide evidence of their accuracy, and as their track records make clear circumstances in which we should have the more confidence in their foretelling, prediction markets may become the stock markets of public policy, prominently covered and displayed as legislators and citizens vote. This development would encourage more discussion of consequences.

Technological change will likely improve prediction markets. Already some theorists are experimenting with combinatorial prediction markets to gather more information.[31] These markets take a bet on a variety of events and use computationally demanding algorithms to approximate changes in prices of bets on related events. For instance, assume that a horse race permits a variety of bets on horses—to win, to place, to show—as well as bets on horses to place in a specific order. The outcomes on those bets are, of course, related. A bet on a horse to show raises the predicted probability that the horse will finish first. With algorithms representing those changes, one can get more information on thinly traded markets, because markets can gain information from related markets. The methodology can be used to create combinatorial markets on related public policy events.

Empirical analysis of past policies will also be aided by simulations of other worlds. These simulations will be of different kinds. Some will slightly change the conditions in which past policies worked. If the effects of policy are largely similar given those changes, the conclusions about the past policies will appear more robust. More robust conclusions about past policies then aid the operation of prediction markets and other assessments of current policy.

Counterfactual analysis of past policies will be a more ambitious enterprise. For instance, one might simulate what would have been the effect of different kinds of economic stimulus packages, including no stimulus at all, on the 2008 recession that President Obama inherited. These kinds of simulations will become more possible as computers capture more complete data about the world at every given instance and as greater computing power then permits fuller evaluation of the consequences of past policy.

Computer games also create alternate worlds. We already have games with tens of thousands of simultaneous players experimenting with virtual realities. As John Holland notes, "Traditionally skills in exploring alternatives have been sharpened via board games, war games, and the like, but that context has always been limited. Video games have broadened that context—games like SimCity and Civilization substantially increase our sensitivity to intricate sociopolitical interactions—and the interfaces are much more realistic, allowing the ordinary citizen to explore options with ease."[32] Games could become more and more similar to the real world as prestige and money become part of their play. Besides revealing information about policy effects, these games may become a method of constraining bias as citizens take roles in the game that are different from those they have in real life. This prospect may be less futuristic than it sounds. Already games are harnessing people to solve scientific problems. Gamers used a program called Foldit to collaborate in predicting the structure of an enzyme that had eluded researchers.[33] Gaming can produce new and useful information.

The evolution of such specific information mechanisms is uncertain. What is certain is that politics needs to continue the progress in information production and use that has been marked out now for two millennia. The so-called Fermi paradox suggests that the alternatives to such progress may be bleak indeed. Given that many billions of solar systems like our own have existed for billions of years, enough time has passed for forms of extraterrestrial intelligence to have developed and spread across the galaxies. Reflecting on the vastness of time in which extraterrestrial life could have reached the earth, the famous physicist Enrico Fermi wondered, "Where are they?" One of the most plausible explanations of the absence of extraterrestrial intelligence is also one of the most frightening. It is intrinsic to the nature of any intelligent life to expand knowledge collectively over time and for its civilization to develop at an accelerating rate. But it is also intrinsic to the nature of intelligent life that its civilization will destroy itself as the rate of change exceeds its capacity to adapt to the challenges that this acceleration produces.

The only way to beat these possibly cosmic odds is to increase our capacity to make wise decisions. From the time Pericles saw democracy as a way to debate the consequences of political action, successful societies have sought better mechanisms of information to assess policy results. Over time society has made use of technology to make its analysis of policy more sophisticated and comprehensive. Today the pace of change is faster, and thus the need to match political to technological change is more urgent. Man is both Homo faber—nature's preeminent maker of tools to change the world—and Homo sapiens—nature's master of symbols and language to represent and understand it. We will continue to thrive only if these capacities develop in tandem.

# Acknowledgments

THIS BOOK HAS ARGUED THAT politics can be improved by using information technology to prompt more minds to evaluate the consequences of policy. The book itself has been improved by the many minds that have probed and refined its arguments.

The Searle Center of Northwestern University convened a roundtable in September 2011 to discuss an earlier draft of the book. Participants were Robert Bennett, Tonja Jacobi, Andrew Koppelman, Jim Lindgren, Jide Nzelibe, Max Schanzenbach, and Jim Pfander, all of Northwestern Law School; Jamie Druckman of the Political Science Department at Northwestern University; Brian Fitzpatrick of Vanderbilt Law School; James Grimmelman of New York Law School; Robin Hanson of the Economics Department of George Mason University; Nelson Lund of George Mason Law School; Michael Rappaport of San Diego Law School; Brad Smith of Capital Law School; and Berin Szoka of TechFreedom. I owe them all a debt for vigorous debate over the ideas offered here.

I am also grateful to five friends—Andrew Clark, Steve Lubet, Mark Movsesian, John Pfaff, and Walter Stahr—who read portions of the manuscript and provided me with invaluable feedback. Conversations with Stephen Altschul, Sonia Arrison, Steve Calabresi, Ryan Calo, Seth Chandler, John Duffy, Neal Devins, Dan Hunter, Gary Marchant, Gillian Metzger, Paul Schwartz, Larry Solum, and Ilya Somin invigorated the project. I profited from presentations of my ideas at the Law, Science, and Innovation Center at Arizona State University; the University of Houston Law School; the Marquette Law School; the Singularity University; Stanford Law School; and the William and Mary Law School. Some years ago Michael Abramowicz taught a course with me on law and accelerating technology—appropriately by video link—and I learned a great deal from our exchanges. His book, *Predictocracy: Market Mechanisms for Public and Private Decision Making*, along with Ray Kurzweil's *The Singularity Is Near: When Humans Transcend Biology*, provided an important catalyst for the ideas developed here.

Essays related to this work have been published as "Age of Empirical," *Policy Review* 137(2006): 61; "Laws for Learning in an Age of Acceleration," *William and Mary Law Review* 53 (2011): 305; "A Politics of Knowledge," *National Affairs* (Winter 2012); and "Accelerating Regulatory Review," in *The Nanotechnology Challenge: Creating Legal Institutions for Uncertain Risks* (Cambridge University Press, 2012). I am grateful to the editors for their suggestions, particularly Yuval Levin of National Affairs.

This book would never have been written but for Chuck Myers of Princeton University Press. Four years ago he came by my office to talk about possible book ideas and encouraged me to pursue this one. Without his intervention, I might have remained content to write articles. He has since provided a sympathetic sounding board for a neophyte book author. And without the introduction of Lee Epstein, a source of encouragement for the enterprise, I would never have met Chuck. My thanks also to Ann Adelman and Jill R. Hughes, excellent and sympathetic copyeditors. And I have been fortunate in hard working research assistants: Jonathan Bringewatt, Noah Brozinsky, Claire Hoffman, Branden Stein, Joshua Steinman, and Matthew Underwood.

My greatest debts are to my parents and to my wife, Ardith Spence. From the very beginning my parents gave me the confidence to try out bold ideas without fear of failure. Ardith's thoughtful suggestions given with loving support have resulted in numerous improvements to the book. In preparing a book about how we need to adapt to world that is rapidly transforming, their essential contributions reminded me of the unchanging value of enduring virtues.

# Appendix

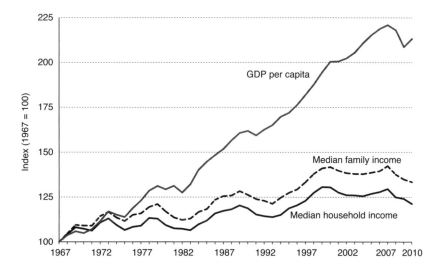

**Figure 1.** United States income growth, 1967–2010. Author's calculations based on data from U.S. Census (median family income and median household income in CPI-U-RS adjusted 2010 dollars) and U.S. Department of Commerce Bureau of Economic Analysis (GDP per capita in 2005 chained dollars). Adapted from "Measuring Economic Well-Being: GDP vs. Median Income" by Anthony Calabrese (http://www.stateoftheusa.org/content/measuring-economic-well-being .php).

# Notes

## Introduction

1. "Petition the White House with We the People," WhiteHouse
.gov, Sept. 22, 2011, http://www.whitehouse.gov/blog/2011/09/22/ petition
-white-house-we-people.
2. Ian Morris, *Why the West Rules—For Now* (New York: Farrar, Straus and
Giroux, 2011), 607.
3. David Warsh, *Knowledge and the Wealth of Nations* (New York: W. W.
Norton, 2006), 288–304.
4. David Deutsch, *The Beginning of Infinity: Explanations That Transform
the World* (New York: Viking, 2011).
5. For the best discussion of this process, see Joel Mokyr, *The Lever of
Riches: Technological Creativity and Economic Progress* (New York: Oxford
University Press, 1992).

## Chapter 1: The Ever Expanding Domain of Computation

1. On the idea of domains in technologies, see Brian Arthur, *The Nature of
Technology: What It Is and How It Evolves* (New York: Free Press, 2009), 61. A
domain forms "a constellation of technologies—a mutually supporting set" (63).
Arthur himself sees the digital technologies of computation as forming a domain
that is drawing other technologies within it (63).
2. "Moore's Law," Intel.com, http://www.intel.com/about/companyinfo
/museum/exhibits/moore.htm.
3. Dan Burke and Mark Lemley, "Policy Levers in Patent Law," *Virginia Law
Review* 89 (2003): 1575, 1620n147.
4. Martin Hilbert and Priscila Lopez, "The World's Technological Capacity
to Store, Communicate, and Compute Information," *Science* 332 (2011): 60–65.
5. Keith Kleiner, "Body 2.0: The Continuous Monitoring of the Human
Body," Singularity Hub, March 20, 2009, http://singularityhub.com/2009/03/20
/body-20-continuous-monitoring-of-the-human-body.
6. Kenrick Vezina, "Stick-On Electronic Tattoos," MIT Technology Review,
Aug. 11, 2011, http://www.technologyreview.com/computing/38296/?p1=A3.
7. Michio Kaku, *Physics of the Future: How Science Will Shape Human
Destiny and Our Daily Lives by the Year 2000* (New York: Doubleday, 2011),
300–301.

8. Ibid.

9. Robert E. Lucas Jr., "On the Mechanics of Economic Development," *Journal of Monetary Economics* 22 (1988): 3.

10. Ray Kurzweil, "Making the World a Billion Times Better," *Washington Post*, April 13, 2008, B-4.

11. Hans Moravec, *Robot: From Mere Machine to Transcendent Mind* (New York: Oxford University Press, 2000), 7.

12. Jeremy Geelan, "'We See No End in Sight,' Says Intel's Pat Gelsinger," *SOA World Magazine*, May 1, 2008, http://java.sys-con.com/read/557154.htm.

13. For discussion of 3-D chips and their extension of Moore's law, see David H. Freedman, "3-D Transistors," *MIT Technology Review*, May/June 2012, http://www.technologyreview.com/article/40243.

14. Ray Kurzweil, *The Singularity Is Near: When Humans Transcend Biology* (New York: Viking, 2005), 67.

15. Ibid.

16. For a good introduction to quantum computing, see George Johnson, *A Shortcut through Time: The Path to the Quantum Computer* (New York: Knopf, 2004).

17. President's Council of Advisors on Science and Technology, "Designing a Digital Future: Federally Funded Research and Development in Networking and Information Technology," Dec. 2010, http://www.whitehouse.gov/sites/default/files/microsites/ostp/pcast-nitrd-report-2010.pdf.

18. Kaku, *Physics of the Future*, 117.

19. Ibid., 6.

20. Mokyr, *The Lever of Riches*, 11.

21. Neal Stephenson, "Innovation Starvation," *World Policy Journal* 28 (2011): 11–16, available at http://www.worldpolicy.org/journal/fall2011/innovation-starvation.

22. Robert Friedel, *A Culture of Improvement: Technology and the Western Millenium* (Cambridge: MIT Press, 2007).

23. Ibid.

24. Martin Weitzman, "Recombinant Growth," *Quarterly Journal of Economics* 113 (1998): 331.

25. John Lauerman, "Complete Genomics Drives Down Cost of Genome Sequence to $5,000," Bloomberg.com, Feb. 5, 2009, http://www.bloomberg.com/apps/news?pid=newsarchive&sid=aEUlnq6ltPpQ. See also Kurzweil, *Singularity*: 206-226.

26. Eric Topol, *The Creative Destruction of Medicine: How the Digital Revolution Will Create Better Health Care* (New York: Basic Books, 2012), 1–7.

27. Ibid., 16.

28. Richard P. Feynman, "There's Plenty of Room at the Bottom," *Caltech Engineering and Science* 23 (1960): 22, available at http://calteches.library.caltech.edu/47/2/1960Bottom.pdf. See also Kurzweil, *Singularity*: 226-259.

29. Deepak Srivastava and Satya N. Atluri, "Computational Nanotechnology: A Perspective," *Computer Modeling in Engineering Sciences* 3 (2002): 531.

30. For a discussion of the need for embodied computation in nanotechnology, see B. J. MacLennan, "Computation and Nanotechnology." *Interna-

*tional Journal of Nanotechnology and Molecular Computation* 1, no. 1 (2009): i–ix.

31. For a recent description of such printers, see "3D-Printer with Nanoprecision," Nanotechnology Now, http://www.nanotech-now.com/news.cgi?story_id=44693.

32. For a description, see "The Printed World," *Economist* (Feb. 10, 2011), http://www.economist.com/node/18114221.

33. "A Third Industrial Revolution," *Economist* (April 21, 2012), http://www.economist.com/node/21552901.

34. Topol, "Creative Destruction of Medicine," 138–40.

35. Gregory Mandel, "Nanotechnology Governance," *Alabama Law Review* 59 (2008): 1323, 1373.

36. Ibid., 1340, n.110.

37. Ramez Naam, "Smaller, Cheaper Faster," *Scientific American Blog*, March 16, 2011, http://www.scientificamerican.com/blog/post.cfm?id=smaller-cheaper-faster-does-moores-2011-03-15.

38. For a description of eSolar, see the company's website at http://www.esolar.com.

39. Robin Hanson, "The Economics of the Singularity," *IEEE Spectrum* 45 (2008): 6, available at http://www.spectrum.ieee.org/jun08/6274.

40. Nicholas Wade, *Before the Dawn: Rediscovering the Lost History of Our Ancestors* (New York: Penguin, 2006), 7.

41. Ibid., 125.

42. David Hounshell, *From the American System to Mass Production, 1800–1932* (Baltimore: Johns Hopkins University Press, 1984), 15–46.

43. Hanson, "Economics of the Singularity."

44. Morris, *Why the West Rules*, 125–35.

45. Ibid., 156 (fig. 3.7).

46. Ibid., 512.

47. Kurzweil, *Singularity*, 35–51.

48. Ibid., 14–20.

49. Ibid., 19.

50. Ray Kurzweil, "The Law of Accelerating Returns," Kurzweil Accelerating Intelligence, March 7, 2001, http://www.kurzweilai.net/the-law-of-accelerating-returns.

51. Maria Tess Shier, "The Way Technology Changes How We Do What We Do," *New Directions for Student Services* (2005): 77–87. See also Matthew Vanden Boogart, "Uncovering the Social Impacts of Facebook on a College Campus," MS thesis, Kansas State University, 2006, available at http://hdl.handle.net/2097/181.

52. "Economic Growth Is Exponential and Accelerating, v2.0," The Futurist, July 12, 2007, http://futurist.typepad.com/my_weblog/2007/07/economic-growth.html.

53. Paula Skokowski, "Data Tsunami: 5 Exabytes of Data Created Every 2 Days?" Secure File Transfer and Collaboration Blog, Aug. 9, 2010, http://www.accellion.com/blog/2010/08/data-tsunami-5-exabytes-of-data-created-every-2-days; Gareth Morgan, "Total Data Created in 2011 to Hit 1.8ZB," June 30,

2011, Computing.Co.UK, http://www.computing.co.uk/ctg/news/2082810/total -created-2011-hit-18zb.

54. Michael McAleer and Daniel Slottje, "A New Measure of Innovation: The Patent Success Ratio," *Scientometrics* 63 (2005): 421, available at http://www .iemss.org/iemss2004/pdf/econometric/mcalasim.pdf.

55. Friedel, *Culture of Improvement*, 275–76.

56. Peter Thiel, "The End of the Future," *National Review* Online, Oct. 3, 2011, http://www.nationalreview.com/articles/278758/end-future-peter-thiel.

57. Tyler Cowen, *The Great Stagnation* (New York: Dutton, 2011).

58. Ibid., 16–22.

59. "Global Poverty: A Fall to Cheer," *Economist*, March 3, 2012, available at http://www.economist.com/node/21548963.

60. David Cay Johnston, "Income Gap Is Widening, Data Shows," *New York Times*, March 29, 2007, http://www.nytimes.com/2007/03/29/business/29tax. html. See also Table P-1, Total CPS Population and Per Capita Income, All Races: 1967 to 2010, http://www.census.gov/hhes/www/income/data/historical/people /index.html (accessed April 25, 2011).

61. Anthony Calabrese, "Measuring Economic Well-Being: GDP vs. Median Income," The State of the USA, July 6, 2010, http://www.stateoftheusa.org/content /measuring-economic-well-being.php.

62. The appendix is drawn from this graph-creating site: http://www .stateoftheusa.org/content/measuring-economic-well-being.php.

63. Sherwin Rosen, "The Economics of Superstars," *American Economics Review* 71 (1981): 845.

64. W. Michael Cox and Richard Alm, *Myths of Rich and Poor* (New York: Basic Books, 1999), 42, 216. See also U.S. Census Bureau, *2005–2009 American Community Survey 5-Year Estimates*.

65. Kate Pickert, "Employer-Based Insurance: Paying More, Getting Less," *Time* Online, Oct. 26, 2009, http://www.time.com/time/nation/article /0,8599,1932184,00.html#ixzz1MpGXUdsX.

66. Cowen, *Great Stagnation*, 30–36.

67. N. Gregory Mankiw, "Beyond Those Health Care Numbers," *New York Times*, Nov. 4, 2007, http://www.nytimes.com/2007/11/04/business/04view.html.

68. Office of Minority Health, "Asian American/Pacific Islander Profile" (2011), http://minorityhealth.hhs.gov/templates/browse.aspx?lvl=2&lvlID=53. Compare with Organization for Economic Co-operation and Development, "Health: Life Expectancy Improving, Spending Rising in Asia-Pacific" (Dec. 21, 2010), http://www.capsentis.com/health/article/Life_expectancy_improving _spending_rising_in_Asia_Pacific_00036_1.html.

69. Steven Reinberg, "Cancer Survival Depends on Where You Live," *Washington Post*, July 17, 2008, http://www.washingtonpost.com/wp-dyn/content/ article/2008/07/16/AR2008071602480.html.

70. "Larry Summers on Debt, Bubbles, and Obama," *Fortune Tech*, CNN *Money*, July 19, 2011, http://tech.fortune.cnn.com/2011/07/19/brainstorm-tech -video-larry-summers-transcript.

71. U.S. Census Bureau, "Family Net Worth: Mean and Median Net Worth in Constant (2007) Dollars by Selected Family Characteristics: 1998 to 2007,"

*Statistical Abstract of the United States: 2011*, table 720; Federal Reserve Bulletin, "Survey of Consumer Finances, 1983: A Second Report," 863, table 7. ("Net worth" is the difference between gross assets and liabilities and is used synonymously with "wealth" in Census Bureau data and Federal Reserve Bulletins.) Data adjusted for inflation using the Bureau of Labor and Statistics Inflation Calculator, http://www.bls.gov/data/inflation_calculator.htm.

72. See Bruce D. Meyer and James X. Sullivan, "Consumption and Income Inequality in the U.S. Since the 1960s," working paper, Oct. 18, 2010, http://harrisschool.uchicago.edu/faculty/web-pages/Inequality60s.pdf.

73. Avi Friedman and David Krawitz, *Peeking through the Keyhole: The Evolution of North American Homes* (Montreal: McGill-Queen's University Press, 2001), xi-xii.

74. Advisory Commission to Study the Consumer Price Index, "The Boskin Commission Report: Toward a More Accurate Measure of the Cost of Living," Dec. 4, 1996, Social Security Online, http://www.ssa.gov/history/reports/boskinrpt.html.

75. Jerry Hausman, "Cellular Telephones, New Products, and the CPI," *Journal of Business and Economic Statistics* 17 (1990): 188.

76. "Boskin Commission Report."

77. Lance Whitney, "Average Net User Now Online 13 Hours per Week," CNET News, Dec. 23, 2009, http://news.cnet.com/8301-1023_3-10421016-93.html. See also Matthieu Pélissié du Rausas, James Manyika, Eric Hazan, Jacques Bughin, Michael Chui, and Rémi Said, "The Net's Sweeping Impact on Growth, Jobs, and Prosperity," McKinsey Global Institute, May 2011, http://www.mckinsey.com/Insights/MGI/Research/Technology_and_Innovation/Internet_matters.

78. For a fuller description of this tale, see "The Legend of the Chessboard," http://britton.disted.camosun.bc.ca/jbchessgrain.htm.

79. Cowen, *Great Stagnation*, 80–83.

80. On religion and radical life extension, see Sonia Arrison, *100+: How the Coming Age of Longevity Will Change Everything from Careers and Relationships to Family and Faith* (New York: Basic Books, 2011), 151–75.

81. C. Eugene Steuerle, "America's Related Fiscal Problems," *Journal of Policy Analysis and Management*, http://www.urban.org/uploadedpdf/1001447-Americas-Related-Fiscal-Problems.pdf.

82. Robert Shiller, "Building a Better Safety Net for Workers," *Japanese Times*, March 27, 2006, http://cowles.econ.yale.edu/news/shiller/rjs_06-03-27_building.htm.

## Chapter 2: Democracy, Consequences, and Social Knowledge

1. David Held, *Models of Democracy* (Oxford: Blackwell, 1996), 14.

2. Richard A. Posner, *Law, Pragmatism, and Democracy* (Cambridge: Harvard University Press, 2003), 18.

3. Michael S. Lewis-Beck et al., *The American Voter Revisited* (Ann Arbor: University of Michigan Press, 2008), 195.

4. The importance of economic growth to American voters is demonstrated in the relationship between economic performance and reelection of presidential incumbents. See, e.g., Ray Fair, "The Effect of Economic Events on Votes for President," *Review of Economics and Statistics* 60 (1978): 159.

5. George W. Bush, "Remarks on the Bipartisan Congressional Tax Relief Agreement and an Exchange with Reporters," American Presidency Project, May 1, 2001, http://www.presidency.ucsb.edu/ws/index.php?pid=45569&st=&st1.

6. Barack Obama, "Statement on Signing the American Recovery and Investment Act of 2009," American Presidency Project, Feb. 17, 2009, http://www.presidency.ucsb.edu/ws/index.php?pid=85782#axzz1r5lNObcZ.

7. Josiah Ober, *Democracy and Knowledge* (Princeton, NJ: Princeton University Press, 2008), 13.

8. John O. McGinnis and Mark L. Movsesian, "The World Trade Constitution," *Harvard Law Review* 114 (2000): 511, 542.

9. Lowell C. Rose et al., "The 29th Annual Phi Delta Kappan/Gallup Poll of the Public's Attitudes toward the Public Schools," *Phi Delta Kappan* 79 (1997): 44.

10. See, e.g., David Barboza, "Shanghai Schools' Approach Pushes Students to Top of Test," *New York Times*, Dec. 29, 2010. The degree of common interest is dependent to some extent on the possibility that the pursuit of a common objective will have advantages for the very substantial majority of citizens. For instance, if a nation can grow in such a manner that the population benefits as a whole in the long run, it is likely to be able to focus more on the common objectives than can a nation with little or no growth, because there the public is likely focus more on divisive distributional issues.

11. Annette Clark, "Abortion and the Pied Piper of Compromise," *NYU Law Review* 68 (1993): 265, 296.

12. Iris Marion Young, "Communication and the Other: Beyond Deliberative Democracy," in *Democracy and Difference: Contesting the Boundaries of the Political*, ed. Seyla Benhabib, 120–21 (Princeton, NJ: Princeton University Press, 1996).

13. Ibid.

14. Michael A. Fitts, "Can Ignorance be Bliss?" *Michigan Law Review* 88 (1990): 917, 944-45.

15. Enrico Giovannini, "Statistics and Politics in a 'Knowledge Society,'" *Social Indicators Research* 86 (2008): 177.

16. The notion that moral sentiments can lead to social beneficial behavior is part of the classical liberal tradition beginning with Adam Smith. See Jerry Z. Muller, *Adam Smith in His Time and Ours: Designing the Decent Society* (Princeton, NJ: Princeton University Press, 1995), 106.

17. See, generally, Mancur Olson, *The Logic of Collective Action* (Cambridge: Harvard University Press, 1971). See also Mancur Olson, *Power and Prosperity: Outgrowing Capitalist and Communist Dictatorships* (New York: Basic, 2000) (contrasting special and encompassing interests).

18. Ithiel de Sola Pool, *Technologies of Freedom* (Cambridge: Harvard University Press, 1984).

19. Terry M. Moe, *Special Interest: Teachers' Unions and America's Public Schools* (Washington, DC: Brookings Institute, 2011).

20. Phillip Nelson, "Political Information," *Journal of Law and Economics* 19 (1976): 321.

21. For example, one-third of farming earnings are due to government agricultural policies, funded by taxpayers, such as price supports. Susanne Lohmann, "An Information Rationale for the Power of Special Interests," *American Political Science Review* 92 (1998): 809. In 1990 agricultural support policies cost the average nonfarm household $1,400 (809). See also Randall R. Rucker and Walter N. Thurman, "The Economic Effects of Supply Controls: The Simple Analytics of the U.S. Peanut Program," *Journal of Law and Economics* 33 (1990): 483.

22. Stephen Coate and Stephen Morris, "On the Form of Transfers to Special Interests," *Journal of Political Economy* 103 (1995): 1210.

23. Burton A. Abrams and Kenneth A. Lewis, "The Effect of Information Costs on Regulatory Outcomes," *Journal of Economics and Business* 39 (1987): 159.

24. Terry M. Moe, "The Internet Will Reduce Teachers Union Power," *Wall Street Journal*, July 18, 2011.

25. Bruce Bimber, *Information and American Democracy: Technology in the Evolution of Political Power* (New York: Cambridge University Press, 2003).

26. Ibid., 33–34.

27. Carolyn L. Funk, "The Dual Influence of Self-Interest and Societal Interest in Public Opinion," *Political Research Quarterly* 53 (2000): 37.

28. Geoffrey Brennan and Loren Lomasky, *Democracy and Decision: The Pure Theory of Electoral Preference* (Cambridge: Cambridge University Press, 1993), 19–53. Americans consider the effect of a policy on the nation as a whole. See Samuel L. Popkin, *The Reasoning Voter: Communication and Persuasion in Presidential Campaigns* (Chicago: University of Chicago Press, 1994), 22.

29. Ober, *Democracy and Knowledge*, 2.

30. Michael Schudson, "The Trouble with Experts–And Why Democracies Need Them," *Theory and Society* 35 (2006): 491.

31. James Surowiecki, *The Wisdom of Crowds* (New York: Anchor, 2005).

32. Anthony Downs, "An Economic Theory of Political Action in a Democracy," *Journal of Political Economy* 65 (1957): 135, 148.

33. Robin Hanson, "Shall We Vote on Values, but Bet on Beliefs?" working paper, George Mason University, 2007, http://hanson.gmu.edu/futarchy.pdf.

## Chapter 3: Experimenting with Democracy

1. Steven I. Friedland, "Law, Social Science, and Malingering," *Arizona State Law Review* 30 (1998): 337, 387.

2. James Lindgren, "Predicting the Future of Empirical Legal Studies," *Boston University Law Review* 86 (2006): 1447, 1449.

3. Ibid.

4. Jonathan Klick and Alexander Tabarrok, "Using Terror Alert Levels to Estimate the Effect of Police on Crime," *Journal of Law and Economics* 48 (2005): 267.

5. Ibid., 271.

6. Ibid., 273-74.

7. J. DiNardo and David S. Lee, "Economic Impacts of New Unionization on Private Sector Employers: 1984–2001," *Quarterly Journal of Economics* 119 (2004): 1383–441.

8. See J. Dinardo, "Natural Experiments and Quasi-Natural Experiments," *New Palgrave Dictionary* Online, http://www.dictionaryofeconomics.com.

9. For a discussion of the huge increase in storage capacity for data, see Alexander Szlay and Jim Gray, "Science in an Exponential World," *Nature* 440 (2006): 413.

10. Declan Butler, "2020 Computing: Everything, Everywhere," *Nature* 440 (2006): 402–405.

11. Peter Wayner, "Life as Captured in Charts and Graphs," *New York Times*, April 20, 2011.

12. Stephen Middlebrook and John Muller, "Thoughts on Bots: The Emerging Law of Electronic Agents," *Business Lawyer* 56 (2000): 341.

13. John Simon, "Wisconsin Influence on Sociological Scholarship," *Law and Social Inquiry* 24 (1999): 143, 189.

14. Lindgren, "Predicting the Future," 1452.

15. Derek Partridge, "A Science of Approximate Computation," http://www.nesc.ac.uk/esi/events/Grand_Challenges/paneld/d17.pdf.

16. It is called the *Journal of Empirical Legal Studies*. See Wiley-Blackwell, http://www.blackwellpublishing.com/journal.asp?ref=1740-1453&site=1.    On the rise of empiricists in law schools, see Henry G. Manne and Joshua Wright, "The Future of Law and Economics: A Discussion," June 2008, George Mason Law and Economics Research Paper No. 08-35, George Mason University School of Law, http://papers.ssrn.com/sol3/papers.cfm?abstract_id=1145421.

17. Barry Eichengreen, "The Last Temptation of Risk," National Interest, March 26, 2010, http://nationalinterest.org/article/the-last-temptation-of-risk-3091.

18. On the importance of peer review for improving empirical scholarship, see Gregory Mitchell, "Empirical Legal Scholarship as Scientific Dialogue," *North Carolina Law Review* 83 (2003): 167.

19. For the 2009 program, see http://weblaw.usc.edu/cels/schedule.cfm.

20. Bruce Cain, "Election Law as a Field," *Loyola of Los Angeles Law Review* 32 (1999): 1105, 1116.

21. Lindgren, "Predicting the Future," 1454; Mitchell, "Empirical Legal Scholarship," 189–94.

22. Michelle M. Mello and Kathryn Zeiler, "Empirical Health Law Scholarship: The State of the Field," *Georgetown Law Journal* 96 (2008): 649, 657.

23. James J. Choi, David Laibson, Brigitte C. Madrian, and Andrew Metrick, "Saving for Retirement on the Path of Least Resistance," in *Behavioral Public Finance*, ed. Edward J. McCaffrey and Joel Slemrod (New York: Russell Sage Foundation, 2006), 304.

24. Stephen Taub, "Rules Finalized for Automatic 401(k)s," CFO.com, Oct. 24, 2007,    http://www.archetype-advisors.com/Images/Archetype/Auto-enrollment/Rules%20Finalized%20for%20Automatic%20401ks.pdf.

25. Leegin Creative Leather Prods. v. PSKS, Inc., 551 U.S. 877 (2007).

26. Margaret E. Slade, "The Effects of Vertical Restraints: An Evidence-Based Approach," in *The Pros and Cons of Vertical Restraints*, ed. Arvid Fredenberg, vol. 7 of Pros and Cons Series (Stockholm: Konkurrensverket, 2008).

27. Joel Mokyr, *The Enlightened Economy: An Economic History of Britain, 1700–1850* (New Haven, CT: Yale University Press, 2009), 40.

28. Ibid., 41.

29. Ibid.

30. Ibid.

31. Pub. L. No. 107-110 (2001).

32. The No Child Left Behind Act mentions scientific evidence more than a hundred times. See Herbert Turner et al., "Populating an International Web-Based Randomized Trials Register in the Social, Behavioral, Criminological, and Education Sciences," *Annals of the American Academy of Political Science* 589 (2003): 205.

33. Pub. L. No. 107-279 (2002).

34. Turner, "Populating a Trials Register," 206.

35. Benjamin Michael Superfine, "New Directions in School Funding and Governance: Moving from Politics to Evidence," *Kentucky Law Review* 98 (2009): 653, 687-88.

36. John O. McGinnis, "Reviving Tocqueville's America: The Rehnquist Court's Jurisprudence of Social Discovery," *California Law Review* 90 (2002): 485.

37. Richard A. Epstein, "Exit Rights under Federalism," *Law and Contemporary Problems* 55 (1992): 147.

38. Henry N. Butler and Jonathan R. Macey, "Externalities and the Matching Principle: The Case for Reallocating Environmental Regulatory Authority," *Yale Law and Policy Review* 14 (1996): 25.

39. Cary Coglianese and Catherine Courcy, "Environmental Regulation," in *The Oxford Handbook of Legal Empirical Research*, ed. Peter Cane and Herbert M. Kritzer (New York: Oxford University Press, 2010), 452.

40. New State Ice Co. v. Liebmann, 285 U.S. 262, 311 (1932) (Brandeis, J., dissenting)

41. Gillian E. Metzger, "Federalism under Obama," *William and Mary Law Review* 53 (2011): 567.

42. Noam N. Levey, "Obama Offers Governors Some Flexibility on Health Law," *L.A. Times*, Feb. 28, 2011, http://articles.latimes.com/2011/feb/28/nation/la-na-governors-healthcare-20110301.

43. Bills requiring committees to make federalism assessments have been introduced in Congress so far without success. Federalism Accountability Act of 1999, S. 1214, 106th Cong. (1999).

44. Under the Bush administration agencies were aggressive about preemption of state law. See Catherine Sharkey, "Preemption by Preamble: Federal Agencies and the Federalization of Tort Law," *DePaul Law Review* 56 (2007): 227.

45. McGinnis, "Reviving Tocqueville's America," 511–16.

46. Peter Linzer, "Why Bother with State Bills of Rights?" *Texas Law Review* 68 (1990): 1573, 1605.

47. A federal constitutional amendment defining marriage as the union between a man and woman is unwise for much the same reasons. See John O.

McGinnis and Nelson Lund, "Lawrence v. Texas and Judicial Hubris," *Michigan Law Review* 102 (2004): 1555, 1613.

48. See, e.g., Andrew Sullivan, *Virtually Normal: An Argument about Homosexuality* (New York: Alfred A. Knopf, 1995), 181–85.

49. Maggie Gallagher, "(How) Will Gay Marriage Weaken Marriage as a Social Institution: A Reply to Andrew Koppelman," *University of St. Thomas Law Journal* 2 (2004): 33.

50. McDonald v. Chicago, 130 S. Ct. 3020, 3109 (2010) (Stevens, J., dissenting); Roth v. United States, 354 U.S. 476, 501–08 (1957) (Harlan, J., concurring in part and dissenting in part).

51. Deborah Merritt, "The Guarantee Clause and State Autonomy: Federalism for a Third Century," *Columbia Law Review* 88 (1988): 1, 3–10.

52. The one exception is First Amendment speech cases. I discuss in chapter 5 why that exception may be consistent with a jurisprudence designed to create the maximum amount of social knowledge.

53. Zelman v. Simmons-Harris, 536 U.S. 639 (2002).

54. This decision has already permitted the continuation of much empirical research on whether vouchers improve school performance. For a description of the research, much of which has been largely favorable to school vouchers, see Patrick J. Wolf, "School Voucher Programs: What the Research Says about Parental School Choice," *BYU Law Review* 2008 (2008): 415.

55. McDonald v. Chicago, 130 S. Ct. 3025 (2010).

56. Some have criticized the constitutional amendment process as too stringent to be workable. I show the amendment process is adequate to the task of constitutional updating in John O. McGinnis and Michael B. Rappaport, *Originalism and the Good Constitution* (Cambridge: Harvard University Press, forthcoming).

57. Thomas H. Davenport, "How to Design Smart Business Experiments," *Harvard Business Review* 87 (Feb. 2009): 68, 71 (2009).

58. Charles Aurther, "Google's Marissa Mayer on the Importance of Real-Time Search," *Guardian*, July 8, 2009, http://www.guardian.co.uk/technology/2009/jul/08/google-search-marissa-mayer.

59. Jim Manzi, "What Social Science Does and Does Not Know," *City Journal* 20 (2010): 14–23, http://www.city-journal.org/2010/20_3_social-science.html.

60. David Greenberg et al., "The Social Experiment Market," *Journal of Economic Perspectives* 13 (1999): 157, 159.

61. Ibid., 160.

62. As recognized by President Obama's economic adviser Larry Summers, see Lawrence H. Summers, "Unemployment," in *The Concise Encyclopedia of Economics*, ed. David R. Henderson (Indianapolis: Liberty Fund, 2008), available at http://www.econlib.org/library/Enc/Unemployment.html.

63. Greenberg, "Social Experiment Market," 160.

64. For a discussion of the latest issues in the school voucher debate, see Terrence Moe, "Beyond the Free Market: The Structure of School Choice," *BYU Law Review* 2008 (2008): 557.

65. Harold J. Krent and Nicholas S. Zeppos, "Monitoring Governmental Disposition of Assets: Fashioning Regulatory Substitutes for Market Controls," *Vanderbilt Law Review* 52 (1999): 1703, 1720.

66. Manzi, "What Social Science Does and Does Not Know," 15.

67. One might well consider the creation of the Congressional Budget Office itself part of a movement toward a politics of learning.

68. Richard Dolinar and S. Luke Leininger, "Pay for Performance or Compliance? A Second Opinion on Medicare Reimbursement," *Indiana Health Law Review* 3 (2006): 397, 406.

69. Of course, there are limits to the permissible scope of randomization. We cannot deprive people of settled rights for the benefit of knowledge, however great. But if the policy options to be randomized are all within the government's authority to provide, and if, further, there is a reasonable basis to believe that all options included are potentially efficacious, the individuals assigned to different programs have no constitutional or moral reason to complain.

70. Stephen F. Fienberg, "Randomization and Social Affairs: The 1970 Draft Lottery," *Science* 171 (1971): 255.

71. Aguayo v. Richardson, 473 F.2d 1090, 1103–08 (2d Cir. 1973) (Friendly, J.). Adam M. Samaha discusses this case extensively in "Randomization in Adjudication," *William and Mary Law Review* 51 (2009): 1, 42–43.

72. Aguayo v. Richardson, 473 F.2d 1093.

73. Ibid., 1109–10.

74. See 5 U.S.C. § 552 (1982), amended by Pub. L. No. 98-620, 98 Stat. 3335 (1984).

75. See Beth Simone Noveck, *Wiki Government* (Washington, DC: Brookings Institution Press, 2009), 107–28.

76. Tim Berners-Lee et al., "The Semantic Web," May 17, 2001, *Scientific American*, http://www.scientificamerican.com/article.cfm?id=the-semantic -web.

77. "Memorandum on Transparency and Open Government," 74 Fed. Reg. 4, 685 (Jan. 26, 2009), http://www.gpo.gov/fdsys/pkg/FR-2009-01-26/pdf/E9-1777.pdf; Peter R. Orszag, "Memorandum for the Heads of Executive Departments and Agencies," Dec. 8, 2009, M-10-06, http://www.whitehouse.gov/omb /assets/memoranda_2010/m10-06.pdf.

78. Ibid.

79. Data.gov, http://www.data.gov.

80. Erica Naone, "The Death of Open Data?" MIT Technology Review, April 19, 2011, http://www.technologyreview.com/web/37420.

81. John Schwartz, "An Effort to Upgrade a Court Archive System to Free and Easy," *New York Times*, Feb. 12, 2009, http://www.nytimes.com/2009/02/13 /us/13records.html.

82. John J. Donahue and Justin Wolfers, "The Ethics and Empirics of Capital Punishment," *Stanford Law Review* 59 (2005): 838.

83. Ibid.

84. The Google settlement, which will affect millions of people, can be seen as kind of a reform of the copyright laws in light of technological change. See Pamela Samuelson, "The Google Book Settlement as Copyright Reform," *Wisconsin Law Review* (2011): 479.

85. David A. Hyman, "Institutional Review Boards: Is This the Least Worst We Can Do?" *Northwestern University Law Review* 101 (2007): 749.

## Chapter 4: Unleashing Prediction Markets

1. The best general book on the subject is Michael Abramowicz, *Predictocracy: Market Mechanisms for Public and Private Decision Making* (New Haven, CT: Yale University Press, 2008).

2. Ober, *Democracy and Knowledge*, 118.

3. Robert W. Hahn, "Statement on Prediction Markets," AEI-Brookings Joint Center for Regulatory Studies, May 2007, Social Science Research Network, http://ssrn.com/abstract=984584.

4. Intrade, http://www.intrade.com/jsp/intrade/contractSearch.

5. See Justin Wolfers and Eric Zitzewitz, "Interpreting Prediction Markets as Probabilities," National Bureau of Economic Research Working Paper No. 12200, 2006, http://bpp.wharton.upenn.edu/jwolfers/Papers/Interpreting PredictionMarketPrices.pdf.

6. A study showed that the Iowa Electronics Future Market outperformed the polls three-quarters of the time. See Joyce E. Berg and Thomas Rietz, "The Iowa Electronic Markets: Styled Facts and Open Issues," in *Information Markets: A New Way of Making Decisions*, ed. Robert Hahn and Paul Tetlock (Washington, DC: American Enterprise Institute Press, 2006), 149–50. Note that the Iowa electronics markets limit bets to five hundred dollars each. Markets with less stringent or no limits should elicit larger investments and more activity, likely improving accuracy.

7. On the theory of prediction markets, see Justin Wolfers and Eric W. Zitzewitz, "Prediction Markets," *Journal of Economic Perspectives* 18 (2004): 107.

8. Michael Ott, "Pope Gregory XIV," in vol. 7 of *The Catholic Encyclopedia* (New York: Robert Appleton Company, 1910), available at http://www.newadvent .org/cathen/07004a.htm.

9. Wolfers and Zitzewitz, "Prediction Markets," 122–23.

10. Ibid.

11. Intrade, http://www.intrade.com/jsp/intrade/contractSearch. At the moment, however, these markets are not thick enough to yield real information.

12. For reasons discussed below, this market did not elicit sufficient funds to be useful, but government policy can help make a thicker market.

13. Justin Wolfers and Eric Zitzewitz, "Five Open Questions about Prediction Markets," San Francisco Federal Reserve Working Paper, 2006-06, http://www .frbsf.org/publications/economics/papers/2006/wp06-06bk.pdf.

14. The advantages of multiplying markets are suggested in M. Todd Henderson et al., "Predicting Crime," *Arizona Law Review* 52 (2010): 15, 44.

15. Ibid., 39–41.

16. Erik Snowberg et al., "Partisan Impacts on the Economy: Evidence from Prediction Markets and Close Elections," *Quarterly Journal of Economics* 122 (2007): 807.

17. Hahn, "Statement on Prediction Markets." But prediction markets will likely also attract people who are not perfectly informed. Such attraction avoids the danger that markets with only perfectly informed rational traders may unravel. See Wolfers and Zitzewitz, "Five Open Questions," 2. Wolfers and Zitze-

witz themselves note various motivations that may depart from rationality and may prompt traders to participate, including the entertainment value and overconfidence (2). The prevalence of the overconfidence bias, particularly among those even with a little knowledge, shows there are some biases that may promote rather than undermine democratic updating.

18. Cass R. Sunstein, "Group Judgments: Statistical Means, Deliberation, and Information Markets," *NYU Law Review* 80 (2005): 962.

19. Cass R. Sunstein, *Infotopia: How Many Minds Produce Knowledge* (New York: Oxford University Press, 2006), 118–21.

20. Lindgren, "Predicting the Future,"1450.

21. Robert S. Erikson and Christopher Wlezien, "Are Political Medical Markets Really Superior to Polls as Election Predictors?"*Public Opinion Quarterly* 72 (2008): 190.

22. Abramowicz, *Predictocracy*, 27.

23. Richard Posner has suggested that nonexperts cannot contribute to some kinds of public policy markets that depend on scientific judgments. I offer reason to disagree in chapter 7.

24. James N. Henslin, *Social Problems* (New York: Prentice Hall, 2011), 503.

25. Popkin, *Reasoning Voter*, 18.

26. Ibid. 80.

27. Herbert A. Simon, "Designing Organizations for an Information-Rich World," in *Computers, Communications, and the Public Interest*, ed. Martin Greenberger (Baltimore: Johns Hopkins University Press, 1971), 37, 40.

28. Popkin, *Reasoning Voter*, 219.

29. For discussion of the importance of broadening such connections as a way of making citizens' choices more information rich, see Popkin, *Reasoning Voter*, 36.

30. Rebecca Haw, Note, "Prediction Markets and Law: A Skeptical Account," *Harvard Law Review* 122 (2009): 1217, 1225, 1228.

31. Ibid., 1228.

32. David Leonhardt, "Prediction Markets and Elections," http://economix. blogs.nytimes.com/2012/03/09/prediction-markets-and-elections.

33. Ibid.

34. Jeffrey Friedman, *No Exit: The Problem with Politics* (forthcoming).

35. John R. Smith, "Poll Methodology Is Key to Accuracy," *Florida Sun Sentinel*, July 21, 2010, http://articles.sun-sentinel.com/2010-07-21/news /sfl-jscol-newcolumn-72110_1_real-poll-candidate-races-voters.

36. Douglas E. Schoen, "Do the Math—Gingrich Is Now the GOP Front Runner," *Fox News*, Nov. 30, 2011, http://www.foxnews.com/opinion/2011/11/30 /do-math-gingrich-is-now-gop-front-runner.

37. Abramowicz, *Predictocracy* 21-22.

38. Erik Snowberg and Justin Wolfers, "Explaining the Favorite-Long Shot Bias: Is it Rick-Love or Misperceptions?" *Journal of Political Economy* 118 (2010): 723.

39. Ibid., 743–44.

40. Markku Kaustia, Eeva Alho, and Vesa Puttonen, "How Much Does Expertise Reduce Behavioral Bias? The Case of Anchoring Effects in Stock Return Estimates," *Financial Management* 37 (2008): 391.

41. Haw, "Prediction Markets," 1235.

42. Amos Tversky and Daniel Kahneman, "Judgment under Uncertainty: Heuristics and Biases," *Science* 185 (1974): 1124; Nicholas Barberis et al., "A Model of Investor Sentiment," *Journal of Finance and Economics* 49 (1998): 307.

43. See Burton G. Malkiel, "The Efficient Market Hypothesis and its Critics," *Journal of Economic Perspectives* 17 (2003): 59, 73–76.

44. Michael Abramowicz, "Information Markets, Administrative Decision-making, and Predictive Cost-Benefit Analysis," *University of Chicago Law Review* 71 (2004): 933, 957. ("In thin markets, trades occur relatively rarely, and there is a danger that the most recent transaction will not represent the market consensus.")

45. Carl Hulse, "Pentagon Abandons Plan for Futures Market on Terror," *New York Times*, July 29, 2003, http://www.nytimes.com/2003/07/29/politics/29WIRE-PENT.html.

46. For examples of prediction market manipulation, see Alexandra Lee Newman, "Manipulation in Political Prediction Markets," *Pepperdine Journal of Business Entrepreneurship and Law* 3 (2010): 205.

47. Abramowicz, *Predictocracy*, 30.

48. Ibid., 31. One exception is markets in which participants can have a direct effect on the outcome to be predicted. Markets in terrorism would be a prime example. See Richard A. Posner, *Catastrophe: Risk and Response* (New York: Oxford University Press, 2004), 175.

49. See, e.g., Robin Hanson and Ryan Oprea, "Manipulators Increase Information Market Accuracy," 2004, http://www.pubchoicesoc.org/papers2005/Hanson_Oprea.pdf.

50. Robin Hanson et al., "Information Aggregation and Manipulation in an Experimental Market," *Journal of Economic Behavior and Organization* 60 (2006): 449–59.

51. Berg and Rietz, "Iowa Electronics Markets," 158.

52. Cf. Donald C. Langevoort, "The Social Construction of Sarbanes-Oxley," *Michigan Law Review* 105 (2007): 1817, 1851.

53. Newman, "Manipulation in Prediction Markets," 229–32.

54. The CFTC has recently requested public comment on this issue. See *Request for Public Comment: Concept Release on the Appropriate Regulatory Treatment of Event Contracts*, 73 Fed. Reg. 25699 (2008).

55. Ibid., 25700.

56. See "CFTC Blocks Political Election 'Gaming,' " http://blogs.market watch.com/election/2012/04/03/cftc-blocks-political-election %E2%80% 98gaming%E2%80%99.

57. 31 U.S.C.A. § 5363(1)-(4) (West 2007).

58. Mark Aubuchon, "The Unlawful Internet Gambling Enforcement Act 2006: A Parlay of Ambiguities and Uncertainties Surrounding the Laws of the Internet Gambling Industry," *Appalachian Journal of Law* 7 (2008): 305, 311.

59. Ibid., 310.

60. Andrea L. Marconi and Brian M. McQuaid, "Betting and Buying: The Legality of Facilitating Financial Payments for Internet Gambling," *Banking Law Journal* 124 (2007): 483, 501–502.

61. William N. Thompson, *Legalized Gambling* (Santa Barbara, CA: ABC-CLIO, 1997), 277.

62. Miriam Cherry and Robert L. Rogers, "Markets for Markets: Origins and Subjects of Information Markets," *Rutgers Law Review* 58 (2006): 339, 355.

63. Tom W. Bell, "Prediction Markets for Promoting the Progress of Science and the Useful Arts," *George Mason Law Review* 14 (2006): 59–60.

64. See, e.g., Lee C. Bollinger, "Journalism Needs Government Help," *Wall Street Journal*, July 14, 2010, A-20.

65. Jacob Gersen, "Legislative Rules Revisited," *University of Chicago Law Review* 74 (2007): 1705, 1711

66. Obama Campaign Statement. ("As president, Obama will not sign any non-emergency bill without giving the American public an opportunity to review and comment on the White House website for five days.") http://change.gov /agenda/ethics_agenda/

67. Katharine Q. Seelye, "White House Changes the Terms of a Campaign Pledge about Posting Bills Online," *New York Times*, June 22, 2009, A-11.

68. H. Res. 5 (2011).

## Chapter 5: Distributing Social Information through Dispersed Media and Campaigns

1. For a list of the very substantial number of specialized blogs by law professors, see "Legal Blogs," Rutgers, http://law-library.rutgers.edu/resources /lawblogs.php.

2. See http://marginalrevolution.com; http://www.calculatedriskblog.com; http://delong.typepad.com; and http://gregmankiw.blogspot.com.

3. See Stephen Mihim, "D.I.Y. Microeconomics, The Year in Ideas; Tenth Anniversary Issue," *New York Times Magazine*, Dec. 19, 2010.

4. Robert Bennett, "Democracy as a Meaningful Conversation," *Constitutional Commentary* 14 (1997): 481, 511.

5. For an empirical discussion of specialized blogs and effects on scholars' reputations, see David Mckenzie and Berk Ozler, "The Impact of Blogs," World Bank Blog, Aug. 10, 2011, http://blogs.worldbank.org/impactevaluations/the-impact-of-blogs-part-ii-blogging-enhances-the-blogger-s-reputation-but-does-it-influence-policy.

6. See, e.g., "Historical Inquiry and Legal Concepts," http://originalismblog. typepad.com/the-originalism-blog/2012/03/historical-inquiry-and-legal-con-ceptsmike-rappaport.html; "The Coming Fall of the New Originalism," http:// www.thefacultylounge.org/2012/03/cornell-guest-post-the-coming-fall-of-the -new-originalism.html. I should mention that Professor Rappaport, the professor defending originalism, and I are frequent coauthors of articles on originalism.

7. See the comments of Mary Dudziak and Orin Kerr to "The Bizzaro World of Originalism and the Rise of the Constitutional Echo Chamber," http://www .thefacultylounge.org/2012/03/the-bizzaro-world-of-originalism-and-the-rise-of -the-constitutional-echo-chamber.html.

8. Philip E. Tetlock, *Expert Political Judgment: How Good Is It? How Can We Know?* (Princeton, NJ: Princeton University Press, 2005), 23.

9. See "Empirical Legal Studies," http://www.elsblog.org/the_empirical_legal_studi.

10. David Deutsch, *The Beginning of Infinity: Explanations That Transform the World* (New York: Viking, 2011), 39.

11. Ibid.

12. Glenn Reynolds, *An Army of Davids* (Nashville: Thomas Nelson, 2006), 89–97.

13. Michael J. Gerhardt, "The Future of the Press in a Time of Managed News," *Florida International University Law Review* 2 (2006–2007): 41, 51.

14. Newspapers themselves now have reporters who focus on bringing the empirical research of academics to the attention of their readers. For an example of the funnel at work, see David Brooks, "The Biggest Issue," *New York Times*, July 29, 2008, A-24.

15. Reynolds, *An Army of Davids*, 125–33. The dispersed media revolution is continuing. Video in the form of YouTube and other clips has only recently begun and yet already has an effect on political campaigns. Vassia Gueorguieva, "Voters, MySpace, and YouTube: The Impact of Alternative Communication Channels on the 2006 Election Cycle and Beyond," *Social Science Computer Review* 26 (2008): 288–300, available at http://ssc.sagepub.com/cgi/rapidpdf/0894439307305636v1. Social networks are also a potential avenue for conveying relevant information, as they permit groups of individuals to easily collaborate on policy initiatives.

16. Popkin, *Reasoning Voter*, 47.

17. Steven Johnson, *Where Good Ideas Come From* (New York: Riverhead Books, 2010), 22.

18. Popkin, *Reasoning Voter*, 91.

19. This view is most associated with Cass Sunstein. See Cass Sunstein, *Republic.com 2.0* (Princeton, NJ: Princeton University Press, 2007).

20. Dan Hunter, "Philippic.com," *California Law Review* 90 (2002): 611, 651. Hunter also casts doubt that theoretical reasons to fear group polarization on the Internet are well founded in the literature of social psychology.

21. Matthew Gentzkow and Jesse M. Shapiro, "Ideological Segregation Online and Offline," National Bureau of Economic Research Working Paper No. 15916, April 2010, National Bureau of Economic Research, http://www.nber.org/papers/w15916.

22. Ibid.

23. Ibid.

24. Jeffrey R. Young, "A Study Finds That Web Users Are More Tolerant Than Non-Users," *Chronicle of Higher Education*, June 15, 2001, http://chronicle.com/article/A.Study-Finds-that-WebUsers/109088.

25. Paul Krugman, "Learning from Europe," *op-ed, New York Times*, Jan. 10, 2010, http://www.nytimes.com/2010/01/11/opinion/11krugman.html.

26. Tino Sanandaji, "Krugman Deceives Yglesias," Super-Economy Blogspot, Jan. 13, 2010, http://super-economy.blogspot.com/2010/01/krugman-deceives-yglesias.html.

27. "Learning from Europe," Greg Mankiw's Blog, Jan. 11, 2010, http://gregmankiw.blogspot.com/2010/01/learning-from-europe.html.

28. Tino Sanandaji, "Dynamic America, Poor Europe," Super-Economy Blogspot, Jan. 12, 2010, http://super-economy.blogspot.com/2010/01/dynamic-america-poor-europe.html.

29. Robert Barro, "The Folly of Subsidizing Unemployment," *Wall Street Journal* Online, Aug. 30, 2010, http://online.wsj.com/article/SB1000142405274 8703959704575454443145772018.html.

30. Joe Weisenthal, "Does Anyone Believe That Unemploment Would Be Just 6.8% if Obama Hadn't Extended Jobless Benefits?" Business Insider, Aug. 30, 2010, http://www.businessinsider.com/robert-barro-on-extending-jobless-benefits-2010-8.

31. E. Tedeschi, "Robert Barro Picks a Hard Fight," Lobster Stuffed with Tacos, Aug. 31, 2010, http://etedeschi.com/2010/08/31/robert-barro-picks-a-hard-fight.

32. For a discussion of partisan media in the early republic, see Marcus Daniel, *Scandal and Civility: Journalism and the Birth of American Democracy* (New York: Oxford University Press, 2008).

33. See, e.g., Alan Wolfe, *Does American Democracy Still Work?* (New Haven, CT: Yale University Press, 2006), 114.

34. William Easterly, *The White Man's Burden: Why the West's Efforts to Aid the Rest Have Done So Much Ill and So Little Good* (New York: Penguin 2006), 15 (citing Charles Lindblom, "The Science of Muddling Through," *Public Administration Review* 19 (1959): 79).

35. Curtis Cartier, "Chrystal Cox, Oregon Blogger, Isn't a Journalist, U.S. Courts Rules," *Seattle Weekly* Blogs, Dec. 6, 2011, http://blogs.seattleweekly.com/dailyweekly/2011/12/crystal_cox_oregon_blogger_isn.php.

36. Stephanie Franzee, "Bloggers as Reporters, An Effect-Based Approach in a New Age of Information Dissemination," *Vanderbilt Journal of Entertainment and Technology Law* 8 (2006): 609, 635.

37. Wikileaks, the website that has disclosed classified information from the U.S. government, has raised this issue anew. See Charles Savage, "After Afghan War Leaks, Revisions in a Shield Bill," *New York Times*, Aug. 4, 2010, A-6.

38. Certain members of Congress sued the Federal Election Commission to get the agency to extend the strictures of the McCain-Feingold Act to the Internet. Shays v. FEC, 337 F. Supp. 2d 28, 65–71 (D.D.C. 2004), affirmed 414 F.3d 76 (D.C. Cir. 2005). So far, however, the FEC has largely exempted the Internet. See Bradley Smith, "The John Roberts Salvage Company: After McConnell, A New Court Looks to Repair the Constitution," *Ohio State Law Journal* 68 (2007): 891, 898.

39. Susan P. Crawford, "The Internet and the Project of Communications Law," *UCLA Law Review* 55 (2007): 359, 393.

40. David Zarefsky, *Lincoln, Douglas, and Slavery: In the Crucible of Public Debate* (Chicago: University of Chicago Press, 1990), 18.

41. Richard Posner, "Unlimited Campaign Spending—a Good Thing?" Becker-Posner Blog, http://www.becker-posner-blog.com/2012/04/unlimited-campaign-spendinga-good-thing-posner.html. In response, Professor Bradley Smith suggests that several of the claims Judge Posner makes in this post are empirically

unsupported. Brad Smith, "Campaign Spending and Voter Knowledge: Posner Posts a Most Unposner-Like Post," http://www.campaignfreedom.org/2012/04/15 /campaign-spending-and-voter-knowledge-posner-posts-an-unposner-like-post.

42. Peter Van Aelst and Knut de Swert, "Politics in the News: Do Campaigns Matter? A Comparison of Political News during Election Periods and Routine Periods in Flanders (Belgium)," *Communications* 34 (2009): 149, 150 (in Western democracies, during campaigns the public pays more attention to issues and candidates, especially how they are presented in the media).

43. Thomas Stratmann, "How Prices Matter in Politics: The Returns to Campaign Advertising," *Public Choice* 140 (Sept. 2009): 357 (finding that advertising during campaigns does impact a candidate's success, and therefore it would be in a candidate's best interest to engage in such advertizing to inform voters).

44. See generally John J. Coleman and Paul F. Manna, "Congressional Campaign Spending and the Quality of Democracy," *Journal of Politics* 62 (2000): 757 passim; Travis N. Ridout et al., "Evaluating Measures of Campaign Advertising Exposure on Political Learning," *Political Behavior* 26 (2004): 201 passim.

45. Xinshu Zhao and Glen L. Bleske, "Measurement Effects in Comparing Voter Learning from Television News and Campaign Advertisements," *Journalism and Mass Communication Quarterly* 72 (1995): 79–80. See also Charles Atkin and Gary Heald, "Effects of Political Advertising," *Public Opinion Quarterly* 40 (1976): 216, 227.

46. Gary C. Jacobson, "The Effects of Campaign Spending in House Elections: New Evidence for Old Arguments," *American Journal of Political Science* 34 (1990): 334, 357.

47. Ibid. See also Stratmann, "How Prices Matter in Politics," 357–77.

48. Chris W. Bonneau and Melinda Gann Hall, *In Defense of Judicial Elections* (New York: Routledge, 2009), 44.

49. Ronald Reagan TV Ad: "It's morning in America again" [video], YouTube, http://www.youtube.com/watch?v=EU-IBF8nwSY.

50. "Sestak Swats Toomey On Social Security" [video], Midterm Election 2010: Campaign Ad Database, Huffington Post, http://www.huffing tonpost.com/2010/09/22/campaign-ads-2010-election_n_683047.html#s156249 &title=Sestak_Swats_Toomey.

51. "NRSC Paints Bennet As Big Spender" [video]. http://www.huff ingtonpost.com/2010/09/22/campaign-ads-2010-election_n_683047 .html#s157479&title=NRSC_Paints_Bennet.

52. Nils-Henrik M. von der Fehr and Kristin Stevik, "Persuasive Advertising and Product Differentiation," *Southern Economics Journal* 65 (1998): 113, 114–15. See also Bibek Banerjee and Subir Bandyopadhyay, "Advertising Competition under Consumer Inertia," *Marketing Science* 22 (2003): 131–32 (citing W. T. H. Koh and H. M. Leung, "Persuasive Advertising," Department of Business Policy, National University of Singapore, 1992).

53. John J. Coleman, "The Distribution of Campaign Spending Benefits across Groups," *Journal of Politics* 63 (2001): 916 passim. See also Vincent Price and John Zaller, "Who Gets the News? Alternative Measures of News Reception

and Their Implications for Research," *Public Opinion Quarterly* 57 (1993): 133 passim.

54. Lauren Elms and Paul M. Sniderman, "Informational Rhythms of Incumbent-Dominated Congressional Elections," in *Capturing Campaign Effects*, ed. Henry Brady and Richard Johnston (Ann Arbor: University of Michigan Press, 2006), 221.

55. Edward J. Lopez, "Up with Campaign Finance," Division of Labour, June 20, 2008, http://divisionoflabour.com/archives/004806.php.

56. Ibid.

57. "2006 Fact Pack, 4th Annual Guide to Advertising Marketing," supplement to *Advertising Age*, Feb. 27, 2006, 10.

58. "Price of Admission," OpenSecrets.org, Center for Responsive Politics, May 2011, http://www.opensecrets.org/bigpicture/stats.php.

59. For another example of complaints about campaign spending, see generally Nicolas Sahuguet and Nicola Persico, "Campaign Spending Regulation in a Model of Redistributive Politics," *Economic Theory* 28 (2006): 119.

60. Federal Election Commission, "Contribution Limits 2011–12," Federal Election Campaign Act, FEC.gov, http://www.fec.gov/pages/brochures/fecfeca .shtml#Contribution_Limits.

61. In a Rasmusssen survey, a majority of those who had an opinion thought that less than ten thousand dollars was not corrupting. Rasmussen Reports, "National Survey of 1,000 Likely Voters Conducted July 8–9, 2009," http://www .rasmussenreports.com/public_content/politics/questions/pt_survey_questions /july_2009/toplines_campaign_finances_july_8_9_2009. The median answer in a Cooperative Congressional Election Study was that ten thousand dollars could be corrupting (draft information provided by Professor Bradley Smith of Capital Law School).

62. See Doe v. Reed, 130 S. Ct. 2811 (2010); Citizens United v. FEC, 130 S. Ct. 876 (2010); McConnell v. FEC, 540 U.S. 93 (2003).

63. Eliza Newlin Carney, "The Deregulated Campaign," Congressional Quarterly, Sept. 17, 2011.

64. Joe Flint, "FCC says TV Stations Must Disclose on Political Ad Spending," *L.A. Times* (April 28, 2012), http://www.latimes.com/business/la-fi-ct-fcc -political-20120428,0,2620299.story.

65. Thomas Stratmann, "Some Talk: Money in Politics. A (Partial) View of the Literature," *Public Choice* 124 (2005): 135, 143. See generally Thomas Stratmann, "Can Special Interests Buy Congressional Votes? Evidence from Financial Services Legislation," *Journal of Law and Economics* 45 (2002): 345, 368. For other examples of arguments that special interest contributions influence political decision making in Congress and therefore results in an overrepresentation of such interests to the detriment of the interests of less organized citizens, see generally Thomas B. Edsall, *Power and Money* (New York: W. W. Norton, 1989); R. Kenneth Godwin, *One Billion Dollars of Influence: The Direct Marketing of Politics* (Chatham, NJ: Chatham House, 1988); Philip M. Stern, *The Best Congress Money Can Buy* (New York: Pantheon, 1988); and Amitai Etzioni, *Capital Corruption: The New Attack on American Democracy* (San Diego: Harcourt, 1984).

66. Joseph E. Cantor, "Campaign Finance," Almanac of Policy Issues, updated Oct. 23, 2002, http://www.policyalmanac.org/government/archive/crs_campaign_finance.shtml.

67. Daniel Houser and Thomas Stratmann, "Selling Favors in the Lab: Experiments on Campaign Finance Reform," *Public Choice* 136 (July 2008); 215-239 passim; Stephen Coate, "Pareto Improving Campaign Finance Policy," *American Economics Review* 94 (2004): 628 passim. See also Stratmann, "Some Talk," 137.

68. Stratmann, "Some Talk," 137.

69. One individual voicing such concern is President Obama, who declared in his State of the Union address one week after the Court's decision that it "will open the floodgates for special interests, including foreign corporations, to spend without limit in our elections. I don't think American elections should be bankrolled by America's most powerful interests." See Robert Barnes, "Reaction Split on Obama's Remark, Alito's Response at State of the Union," *Washington Post*, Jan. 29, 2010, http://www.washingtonpost.com/wp-dyn/content/article/2010/01/28/AR2010012802893.html.

70. Prohibitions on corporate contributions have been in place since the Tillman Act of 1907. See Ch. 420, 34 Stat. 864–65. Congress extended the ban on corporate donations to labor organizations with the War Labor Disputes Act of 1943 (ch. 144, § 9, 57 Stat. 167–68). The Taft-Hartley Act of 1947 (§ 304, 61 Stat. 159–60) made this extension permanent. Limitations on corporate contributions have been upheld by the Court in Buckley v. Valeo and then Citizens United v. FEC. For a good overview of the limitations over the years on corporate contributions, see Federal Election Commission v. Wisconsin Right to Life, Inc., 551 U.S. 449, 507–19 (2007).

71. See 79 A.L.R.3d 491, *Power of Corporation to Make Political Contribution or Expenditure Under State Law*.

72. For a discussion of some of the evidence, see John O. McGinnis, "The Manufactured Hysteria Over Citizens United," SCOTUS report, July 12, 2012, http://www.scotusreport.com/2012/07/12/the-manufactured-hysteria-over-citizens-united/.

73. For an example of this argument see Eric Gomez, "The Erosion of Democracy," Democracy Matters, March 25, 2011, http://www.democracymatters.org/index.php?option=com_content&view=article&id=368:the-erosion-of-democracy&catid=41:dm-in-the-news&Itemid=63.

74. See 2008 National Presidential Exit Polls, available at CNN.com, http://www.cnn.com/ELECTION/2008/results/polls/#USP00p1. See also R. M. Schneiderman, "How Did Rich People Vote, And Why?" *New York Times* Business Day Economix Blog, Nov. 11, 2008, http://economix.blogs.nytimes.com/2008/11/11/how-did-rich-people-vote-and-why.

75. It is almost universally accepted that Hollywood is predominantly left-leaning. For an example of one of the many sources asserting Hollywood is overwhelmingly liberal, see Ben Shapiro, *Primetime Propaganda: The True Hollywood Story of How the Left Took Over Your TV* (Boston: Beacon Broadside, 2011). For evidence that journalists are generally left-leaning, see "The American Journalist: Politics and Party Affiliation," Journalism.org, Oct. 6, 2006, http://www.journalism.org/node/2304. Finally, there is long-standing evidence that academics have traditionally affiliated themselves with the Democratic Party. Two surveys

in 1959 and 1970 of political scientists revealed that 75 percent of respondents were Democrats, far greater than the general population. See Henry A. Turner and Carl C. Hetrick, "Political Activities and Party Affiliations of American Political Scientists," *Western Political Quarterly* 25 (1972): 361, 362.

76. Jacob Rowbottom, "How Campaign Finance Laws Made the British Press so Powerful," *New Republic*, July 25, 2011, http://www.tnr.com/article /world/92507/campaign-finance-united-kingdom-news-corporation.

77. Richard Posner, "Justice Breyer Throws Down the Gauntlet," *Yale Law Journal* 115 (2006): 1699, 1705.

78. David Rosenberg, *Broadening the Base: The Case for a New Federal Tax Credit for Political Contributions* (Washington, DC: American Enterprise Institute, 2002), 7.

79. Ibid.

80. Ibid., 10.

## Chapter 6: Accelerating AI

1. For a definition of Friendly AI, see Singularity Institute for Artificial Intelligence, "Creating Friendly AI," § 1, http://www.singinst.org/upload/CFAI.html, which summarizes the goals of Friendly AI as assuring that AI seeks the elimination of involuntary pain, death, coercion, and stupidity. I might suggest an even weaker definition as simply assuring that AI does not create harm to humans or limit their freedom through either malevolence or stupidity.

2. *2001: A Space Odyssey*, directed by Stanley Kubrick, Metro-Goldwyn-Mayer, 1968; *Wall-E*, directed by Andrew Stanton, Pixar Animation Studios, 2008.

3. Consideration of artificial intelligence has not bulked large public policy. One interesting article analyzes whether artificial intelligence can play the role of a trustee. See Lawrence B. Solum, "Legal Personhood for Artificial Intelligences," *North Carolina Law Review* 70 (1992): 1231.

4. See, e.g., Jules Verne, *From the Earth to the Moon* (New York: Bantam, 1967).

5. Irving John Good, "Speculations Concerning the First Ultraintelligent Machine," *Advances in Computers* 6 (1965): 31–36.

6. Henry Brighton and Howard Selina, *Introducing Artificial Intelligence* (London: Totem, 2003), 42.

7. Allen Newell and Herbert A. Simon, "Computer Science as Empirical Inquiry: Symbols and Search," *Communications of the ACM* 19 (1976): 113, 118.

8. John R. Searle, "Minds, Brains, and Programs," *Behavioral and Brain Science* 3 (1980): 417–18.

9. Ibid., 418.

10. Ibid.

11. See, e.g., Daniel C. Dennett, *Consciousness Explained* (New York: Little Brown, 1991), 439.

12. Kurzweil, *Singularity*, 125–27.

13. David J. Chalmers, "The Singularity: A Philosophical Analysis," *Journal of Consciousness Studies* 17 (2010): 7, available at http://consc.net/papers/singularity .pdf.

14. See Brighton and Selina, *Introducing Artificial Intelligence*, 23.

15. See "Deep Blue Overview," IBM Research, http://www.research.ibm.com /deepblue/watch/html/c.shtml.

16. Fen-hsiung Hsu, "Cracking Go," IEEE Spectrum, Oct. 2007, http://spectrum .ieee.org/computing/software/cracking-go/0.

17. Ibid.

18. See "Data, Data Everywhere," *Economist*, Feb. 27, 2010, 3–5, available at http://www.economist.com/specialreports/displaystory.cfm?story_id=15557443.

19. IBM Watson, http://www-03.ibm.com/innovation/us/watson/index.html.

20. "CMU Robot Car First in DARPA Urban Challenge," SPIE.org, Nov. 5, 2007, http://spie.org/x17538.xml.

21. Stephen Edelstein, "Nevada DMV Announces Regulations for Self-Driving Cars," DigitalTrends.com, Feb. 20, 2012, http://www.digitaltrends.com /cars/nevada-dmv-announces-regulations-for-self-driving-cars.

22. Steve Lohr, "In Case You Wondered, a Real Human Wrote This Column," *New York Times*, Sept. 11, 2011, available at http://www.nytimes. com/2011/09/11/business/computer-generated-articles-are-gaining-traction. html?pagewanted=all.

23. Steven Levy, "The AI Revolution Is On," *Wired*, Dec. 27, 2010, http:// www.wired.com/magazine/2010/12/ff_ai_essay_airevolution.

24. Ibid.

25. "Mouse Brain Simulated on Computer," BBC News, April 27, 2011, http://news.bbc.co.uk/2/hi/6600965.stm.

26. Kurzweil, *Singularity*, 160–67.

27. John Markoff, "The Coming Superbrain," *New York Times*, May 24, 2009, 1, available at http://www.nytimes.com/2009/05/24/weekinreview/24markoff .html.

28. Ibid.

29. John Markoff, "Scientists Worry Machines May Outsmart Man," *New York Times*, July 25, 2009, A-1, available at http://www.nytimes.com/2009/07/26 /science/26robot.html; "Predator Drones and Unmanned Aerial Vehicles (UAVs)," *NewYorkTimes.com*, updated March 20, 2012, http://topics.nytimes.com/top /reference/timestopics/subjects/u/unmanned_aerial_vehicles/index.html.

30. Kurzweil, *Singularity*, 122, 320–30.

31. Bill Joy, "Why the Future Doesn't Need Us," *Wired* 8, no. 4, April 2000, 238, available at http://www.wired.com/wired/archive/8.04/joy.html. Joy is not alone in his concern. See Kevin Warwick, *March of the Machines: The Breakthrough in Artificial Intelligence* (Champaign: University of Illinois Press, 2004), 280–303.

32. Richard Holmes, *The Age of Wonder: How the Romantic Generation Discovered the Beauty and Terror of Science* (New York: Vintage Books, 2010), 94n.

33. Friedel, *Culture of Improvement*, 85, 113, 131, 374.

34. Noah Shachtman, "DARPA Chief Speaks," Danger Room Blog, Wired. com, Feb. 20, 2007, http://blog.wired.com/defense/2007/02/tony_tether_has_1. html.

35. P. W. Singer, *Wired for War: The Robotics Revolution and Conflict in the 21st Century* (New York: Penguin, 2009), 35.

36. Rowan Scarborough, "Unmanned Warfare," *Washington Times*, May 8, 2005, A-1.

37. Vietnam Veterans of America, Quad Cities Chapter, "The Last Fighter Pilot Has Already Been Born," Shoulder to Shoulder, May 16, 2009, http://qcvva299.org/2009/05/16/the-last-fighter-pilot-has-already-been-born.

38. For a popular account in a major film, see *The Hurt Locker*, directed by Kathryn Bigelow, First Light Production, 2009.

39. Singer, *Wired for War*, 111.

40. Ibid., 112.

41. John O. McGinnis and Ilya Somin, "Should International Law Be Part of Our Law?"*Stanford Law Review* 59 (2007): 1175, 1236–38.

42. Gustavo R. Zlauvinen, "Nuclear Non-Proliferation and Unique Issues of Compliance," *ILSA Journal of International and Comparative Law* 12 (2006): 593, 595.

43. William J. Perry, "Proliferation on the Peninsula; Five North Korean Nuclear Crises," *Annals of the American Academy of Politics and Social Science* 607 (2006): 78–79.

44. Ibid.

45. Joy, "Why the Future Doesn't Need Us."

46. "Friendly AI," § 2.

47. Ibid.

48. Robert Wright, *The Moral Animal: Evolutionary Psychology and Everyday Life* (New York: Vintage, 1994), 336–37. See also John O. McGinnis, "The Human Constitution and Constitutive Law: A Prolegomenon," *Journal of Contemporary Legal Issues* 8 (1997): 211, 213.

49. Ibid.

50. "Friendly AI," § 3.2.3.

51. Steve Omohundro, "The Basic AI Drives," http://steveomohundro.com /scientific-contributions.

52. Jonathan H. Adler, "Eyes on a Climate Prize: Rewarding Energy Innovation to Achieve Climate Stabilization," *Harvard Environmental Law Review* 35, no. 1 (2011), available at http://ssrn.com/abstract=1576699; William A. Masters and Benoit Delbecq, "Accelerating Innovation with Prize Rewards," International Food Policy Research Institute Discussion Paper No. 835, Dec. 2008, 10, available at http://www.ifpri.org/sites/default/files/publications/ifpridp00835.pdf.

53. This use of peer review rather than regulation to address technical issues that are not possible to capture through bureaucratic mandates is something that should be considered more generally in the regulatory process. See Stephen P. Croley, "The Administrative Procedure Act and Regulatory Reform: A Reconciliation," *Administrative Law Journal* 10 (1996): 35, 47.

54. Thus the policy recommendations here differ somewhat from the Singularity Institute. It considers relinquishment infeasible. See "Friendly AI", § 4.2.2. But

the institute seems to suggest a hands-off approach by the government (§ 4.2.1). While I agree that regulatory prohibitions are imprudent, government support for Friendly AI is useful to give it the momentum to triumph over potentially less friendly versions, the possibility of which cannot be ruled out a priori. AI is potentially friendly but not necessarily so. Moreover, as I discuss later, the positive externalities of AI for collective decision making suggest that such support is warranted to create information about public policy that otherwise would be underproduced.

55. Cf. Kenneth Anderson, "Rise of the Drones: Unmanned Systems and the Future of War," Written Testimony Submitted to Subcommittee on National Security and Foreign Affairs, Committee on Oversight and Government Reform, U.S. House of Representatives, Subcommittee Hearing, March 23, 2010, 111th Cong., 2nd sess. 2010, available at http://digitalcommons.wcl.american.edu/cgi/viewcontent.cgi?article=1002&context=pub_disc_cong.

56. Kenneth Anderson, "Why Targeted Killing? And Why Is Robotics so Crucial an Issue in Targeted Killing?" Kenneth Anderson's Law of War and Just War Theory Blog, March 27, 2009, http://kennethandersonlawofwar.blogspot.com/2009/03/why-targeted-killing-and-why-is.html#links.

57. "Friendly AI," §§ 1, 2, 4.

58. Atul Gawande, *The Checklist Manifesto: How to Get Things Right* (New York: Metropolitan Books, 2010). Gawande's solution in the medical field is to create checklists focusing on the most important protocols for saving lives (137–38). While a checklist approach may indeed be the best solution to information overload given present technology, one problem with a checklist generally is that it creates a one-size-fits-all approach that is frozen in time. In contrast, AI would allow one to decide what is most important for a particular patient, taking in up-to-date information.

59. Christopher Drew, "Drone Flights Leave Military Awash in Data," *New York Times*, Jan. 10, 2010, A-1, available at http://www.nytimes.com/2010/01/11/business/11drone.html.

60. Stephen H. Muggleton, "Exceeding Human Limits," *Nature* 440 (2006): 409.

61. Ian Ayres, *Super Crunchers: Why Thinking-by-Numbers Is the New Way to Be Smart* (New York: Bantam, 2008), 13.

62. Garry Kasparov, "Garry Kasparov on 'Chess Metaphors': The Chess Master and the Computer," *New York Review of Books*, Feb. 11, 2010, available at http://www.huffingtonpost.com/2010/01/22/gary-kasparov-on-chess-me_n_432043.html.

63. The fount of these algorithms is Nick Littlestone and Manfred Warmuth, "The Weighted Majority Algorithm," 256-61, 30th Annual Symposium on Foundations of Computer Science, 1989, available at http://www.cc.gatech.edu/~ninamf/LGO10/wm.pdf.

64. Joshua M. Epstein and Robert L. Axtell, *Growing Artificial Societies: Social Science from the Bottom Up* (Cambridge: MIT Press, 1996). Social scientists already do this type of data varying in the social network arena, in fact. See Stephen P. Borgatti, "Identifying Sets of Key Players in a Social Network," *Computational and Mathematical Organizational Theory* 12 (2006): 21, 33.

65. David S. Law, "Constitutions," in *The Oxford Handbook of Empirical Legal Research*, ed. Peter Kane and Herbert Kritzer (New York: Oxford University Press, 2011), 391.

66. Ibid.

67. Florin Diacu, *Megadisasters: The Science of Predicting the Next Catastrophe* (Princeton, NJ: Princeton University Press, 2009).

68. Nassim Nicholas Taleb, *Fooled by Randomness: The Hidden Role of Chance in Life and in the Markets* (New York: Random House, 2008).

## Chapter 7: Regulation in an Age of Technological Acceleration

1. James M. Landis, *The Administrative Process* (New Haven, CT: Yale University Press, 1938), 33–35.

2. Ibid., 7. For an excellent discussion of the relation of technology and the administrative state, see John F. Duffy, "The FCC and the Patent System: Progressive Ideals, Jacksonian Realism, and the Technology of Regulation," *University of Colorado Law Review* 71 (2001): 1071.

3. For the federalism executive order, see Executive Order 13, 132 (1999).

4. For Michael Abramowicz's somewhat different focus on regulatory prediction markets, see Abramowicz, *Predictocracy*: 162-194.

5. As suggested by Freeman Dyson, "The Question of Global Warming," *New York Review of Books*, June 12, 2008, http://www.nybooks.com/articles /archives/2008/jun/12/the-question-of-global-warming.

6. Posner, *Catastrophe*, 157.

7. David Leppik, "Photovoltaics Follow Moore's Law," Technocrat, May 16, 2008, http://technocrat.net/d/2008/5/16/41454.

8. Lynn E. Blais and Wendy E. Wagner, "Emerging Science, Adaptive Regulation, and the Problem of Rulemaking Ruts," *Texas Law Review* 86 (2008): 1701, 1731-32.

9. Ibid.

10. Einer R. Elhauge, "Does Interest Group Theory Justify More Intrusive Judicial Review?" *Yale Law Journal* 101 (1991): 31, 43.

11. Adler, "Eyes on a Climate Prize," 17.

12. Ibid., 14

13. X Prize Foundation, http://www.xprize.org.

14. Andrew C. Kadak, "The Societal Fairness of Intergenerational Equity," *Proceedings of the International Conference on Probabilistic Safety Assessment and Management* 2 (1998): 1005–1010.

15. Even without accelerating technology, Neil H. Buchanan has noted that on relatively pessimistic assumptions of economic growth subsequent generations are likely to be much better off. See Neil H. Buchanan, "What Do We Owe Future Generations?" *George Washington Law Review* 77 (2009): 1237, 1272.

16. Cf. Kurzweil, *Singularity*, 95.

17. Neil Buchanan makes the important related point that our generation should try to avoid economic and political crises. The self-interest of this genera-

tion and that of subsequent generations substantially converge in these matters. See Buchanan, "What Do We Owe?" 1284–86.

18. Posner, *Catastrophe*, 184–86.

19. See, e.g., ibid.

20. For discussion of these trends, see Arrison, *100+*, 21–49.

21. I notice that Richard Posner makes a similar analogy, but largely to dismiss it. See Posner, *Catastrophe*, 149.

22. The theory behind radical life extension is that in an age when medical advances occur with increasing rapidity, an individual merely has to live until medical advances permit him to surmount diseases that would otherwise kill him. If he continues to do this, he will have reached an "escape velocity" that will permit a life of indefinite length.

23. Posner, *Catastrophe*, 175–76.

24. Ibid., 120–21.


## Chapter 8: Bias and Democracy

1. Alinda Friedman and Norman R. Brown, "Updating Geographical Knowledge: Principles of Coherence and Inertia," *Journal of Experimental Psychology: Learning, Memory, and Cognition* 26 (2000): 900; Robin Hogarth and Hillel Einhorn, "Order Effects in Belief Updating: The Belief-Adjustment Model," *Cognitive Psychology* 24 (1992): 1, 2.

2. Charles A. O'Reilly III, "Variations in Decision Makers' Use of Information Sources: The Impact of Quality and Accessibility of Information," *Academy of Management Journal* 25 (1982): 756; Avner M. Porat, "Information Effects on Decision-Making," *Behavioral Science* 14 (1969): 98.

3. Leda Cosmides and John Tooby, "Are Humans Good Intuitive Statisticians after All? Rethinking Some Conclusions from the Literature on Judgment under Uncertainty," *Cognition* 58 (1996): 11. ("If making accurate judgments under uncertainty is an important adaptive problem, why would natural selection [not have designed] an accurate calculus of probability?")

4. Yael Niv et al., "Evolution of Reinforcement Learning in Uncertain Environments: A Simple Explanation for Complex Foraging Behaviors," *Adaptive Behavior* 10 (2002): 5–6 (discussing how bees change their behavior to better achieve objectives after getting information from the environment).

5. Timur Kuran, *Private Truths, Public Lies: The Social Consequences of Preference Falsification* (Cambridge: Harvard University Press, 1995).

6. Ibid., 71–73.

7. Bryan Caplan, *The Myth of the Rational Voter: Why Democracies Choose Bad Policies* (Princeton, NJ: Princeton University Press, 2007), 50–51, 78–79.

8. Lewis-Beck et al., *American Voter Revisited*, 145.

9. Richard Posner, *Aging and Old Age* (Chicago: University of Chicago Press, 1995), 53.

10. See, e.g, Dan Kahan, Donald Braman, and James Grimmelman, "Modeling Cultural Cognition," *Social Justice Research* 18 (2005): 283; Dan M. Kahan, "The Cognitively Illiberal State," *Stanford Law Review* 60 (2007): 115.

11. Ibid., 120.

12. Milton Lodge and Charles Taber, "The Rationalizing Voter: Unconscious Thought in Political Information Processing," Dec. 21, 2007, unpublished paper, available at http://papers.ssrn.com/sol3/papers.cfm?abstract_id=1077972.

13. Charles S. Taber and Milton Lodge, "Motivational Skepticism in the Evaluation of Political Beliefs," *American Journal of Political Science* 50 (2006): 755.

14. Dan Kahan, Donald Braman and James Grimmelman, "Modeling Facts, Culture, and Cognition in the Gun Debate," *Social Justice Research* 18 (2005): 283, 293, http://digitalcommons.law.yale.edu/cgi/viewcontent.cgi?article=1104&context=fss_papers.

15. James N. Druckman, "On the Limits of Framing Effects: Who Can Frame?" *Journal of Politics* 63 (2001): 1041, 1042.

16. Steven Pinker, "Mind Games: Does Language Frame Politics?" *Public Policy Research* 14 (2007): 59, 61.

17. See, e.g., James N. Druckman and Kjersten R. Nelson, "Framing and Deliberation: How Citizens' Conversations Limit Elite Influence," *American Journal of Political Science* 47 (2003): 729; Druckman, "On the Limits," 1041, 1042.

18. The classic statement of the power of the median voter in this respect is Anthony Downs, *An Economic Theory of Democracy* (New York: Harper, 1957), 297. The power of the median is felt mostly in the long run, because special interests have differential influence and may retard the policy change desired by the median voters in areas in which they are interested. See Mancur Olson, *The Logic of Collective Action* (Cambridge: Harvard University Press, 1964), 141–44. Thus, teachers' unions can retard movement to school choice, even if the median voter desires it. The rise in empiricism will not dissolve the problem of special interests, but it should enable policy change to move faster, because ignorance of the value of a policy change can be very helpful to special interests.

19. Michael S. Kang, "De-Rigging Elections: Direct Democracy and the Future of Redistricting Reform," *Washington University Law Review* 84 (2006): 667, 709.

20. Samuel Issacharoff, "The Endangered Center: Collateral Damage in American Politics," *William and Mary Law Review* 46 (2004): 415, 427–28.

21. Ordinary legislation is passed under a relatively mild supermajority rule. See John O. McGinnis and Michael B. Rappaport, "Our Supermajoritarian Constitution," *Texas Law Review* 80 (2002): 703, 769–75.

22. See Senate Rule 22.

23. Robert Bennett, "Counter-Conversationalism and the Sense of Difficulty," *Northwestern University Law Review* 95 (2001): 845, 873.

24. Samuel Issacharoff and Richard H. Pildes, "Politics as Markets: Partisan Lockups of the Democratic Process," *Stanford Law Review* 50 (1998): 643, 674–75.

25. Bimber, *Information and American Democracy*, 42.

26. Alexander Hamilton et al., "Federalist No. 10," in *The Federalist Papers*, ed. Clinton Rossiter (New York: Signet, 1999), 45.

27. Bimber, *Information and American Democracy*, 43.

28. For a strong version of this thesis, see Alberto Alesina and Howard Rosenthal, *Partisan Politics, Divided Government, and the Economy* (New York: Cambridge University Press, 1995), 2.

29. Ibid.

30. For partisan composition of government over the last sixty two years, see http://en.wikipedia.org/wiki/Divided_government.

31. The so-called securing the base strategy has received a lot of attention in the press; see, e.g., Ronald Brownstein, "Bush Aims to Solidify His Base," *L.A. Times*, Aug. 22, 2004, http://articles.latimes.com/2004/aug/22/nation/na-strategy 22. It is well analyzed in the legal literature as well. Samuel Issacharoff, "Private Parties with Public Purposes: Political Parties, Associational Freedoms, and Partisan Competition," *Columbia Law Review* 101 (2001): 274, 306–307.

32. Many political scientists agree that strong partisan attitudes do not remain stable over time, but rather change in response to exposure to new information. See Alan Gerber and Donald P. Green, "Rational Learning and Partisan Attitudes," *American Journal of Political Science* 42 (1998): 794. That new information changes voter behavior, including turnout. See John G. Matsusaka, "Explaining Voter Turnout Patterns: An Information Theory," *Public Choice* 84 (1995): 91.

33. Ilya Somin, "Voter Ignorance and the Democratic Ideal," *Critical Review* 12 (1998): 413, 431. Professor Somin raises the concern that the better informed will not be representative of the less informed, raising concerns about their undue influence. But here I am considering improving democratic updating about facts, where we are concerned about accuracy, rather than aggregating preferences, where we would be concerned about representativeness. Experts and those who bet on prediction markets are also likely to be unrepresentative, but they can nevertheless make policies better by helping democracy get the facts right.

34. For an example, see Larry Bartels, "The American Public's Defense Spending Preferences in the Post-Cold War Era," *Public Opinion Quarterly* 58 (1994): 479.

35. Donald R. Kinder and Lynn M. Sanders, "Mimicking Political Debate with Survey Questions: The Case of White Opinion on Affirmative Action for Blacks," *Social Cognition* 8 (1990): 73, 90. But see Thomas E. Nelson et al., "Toward a Psychology of Framing Effects," *Political Behavior* 19 (1997): 221, 234.

36. James N. Druckman, "The Politics of Motivation," unpublished paper, June 22, 2011, available at http://faculty.wcas.northwestern.edu/~jnd260 /publications.html.

37. Aaron B. Strauss, "Political Ground Truth: How Personal Issue Experience Counters Partisan Bias," PhD diss., Princeton University, Sept. 2009, http:// www.mindlessphilosopher.net/progress/strauss_dissertation_pre-library.pdf.

38. Druckman, "Politics of Motivation."

39. Druckman, "On the Limits," 1041, 1052.

40. "Inside Obama's Sweeping Victory," Pew Research Center Publications, Nov. 5, 2008, http://pewresearch.org/pubs/1023/exit-poll-analysis-2008.

41. See, e.g., Jeffrey J. Rachlinksi, "Cognitive Errors, Individual Differences, and Paternalism," *University of Chicago Law Review* 73 (2006): 207, 216–24.

42. Ibid., 217. Some researchers in fact see biases as in larger measure driven by cognitive errors. See Keith Stanovich and Richard F. West, "Individual Differences in Reasoning," *Behavioral and Brain Sciences* 23 (2000): 645.

43. Gary C. Jacobson, "Perception, Memory, and Partisan Polarization on the Iraq War," *Political Science Quarterly* 125 (2010): 31, 36. For discussion of the view that the vast majority of people in the United States are not polarized between partisan positions, see Morris Fiorina, *Culture War? The Myth of a Polarized America* (New York: Pearson Education, 2006), 33–57; Alan Wolfe, *One Nation After All* (New York: Viking, 1998), 88–133.

44. Robert Y. Shapiro and Yaeli Bloch-Elkon, "Do the Facts Speak for Themselves? Partisan Disagreement as a Challenge to Democratic Competence," *Critical Review* 20 (2008): 115, 123.

45. Morris P. Fiorina and Samuel J. Abrams, "Political Polarization in the American Public," *Annual Review of Political Science* 11 (2008): 563.

46. Jacobson, "Perception, Memory," 33.

47. Ibid., 39.

48. Dan Balz and Jon Cohen, "Independent Voters Favor Democrats by 2 to 1 in Poll," *Washington Post*, Oct. 24, 2006, A-1.

49. "Exit Polls: Bush, Iraq Key to Outcome," CNN.com, Nov. 8, 2006, www .cnn.com/2006/POLITICS/11/08/election.why/index.html.

50. Professor Kahan and his co-authors, Professor Donald Braman and James Grimmelman, divide the world into quadrants, "each of which represents a distinct cultural orientation." Kahan et al., "Modeling Facts," 283, 290. People so divided have difficulty updating on the basis of information, because in fixing their beliefs, they consider only the beliefs of those in their quadrant. As a result, society does not update, because people follow those with cognitive biases similar to their own. This model does not reflect the world if people are arrayed along a spectrum or if they interact in their lives with other of different ideological views.

51. Paul A. Klaczynski and Billi Robinson, "Personal Theories, Intellectual Ability, and Epistemological Beliefs: Adult Age Differences in Everyday Reasoning Biases," *Psychology and Aging*, 15 (2000): 400, 411–13. If individuals make tradeoffs between adhering to worldviews and updating to reflect new information (see Caplan, *Myth of the Rational Voter*, 17), it may be rational for older people to be slower to update. They have less time to reap the benefits of a political world built on accurate information.

52. For a dramatic graph of the majority support by those under thirty for same-sex marriage in states, including conservative states, see "Future Trends for Same-Sex Marriage Support?" AndrewGelman.com, June 21, 2009, http://www .stat.columbia.edu/~cook/movabletype/archives/2009/06/future_trends_f_1. html.

53. Nancy J. Knauer, "The Recognition of Same-Sex Relationships: Comparative Institutional Analysis, Contested Goals, and Strategic Institutional Choice," *University of Hawaii Law Review* 28 (2005): 23, 39. See William Saletan, "Original Skin: Blacks, Gays and Immutability," Slate, Nov. 13, 2008, http://www.slate. com/articles/health_and_science/human_nature/2008/11/original_skin.html.

54. This generational change also casts doubt on the ability of worldviews to ultimately obstruct change. Presumably there are still individuals who are egalitarian, hierarchical, or individualist in similar proportions in each generation. Yet what is accepted in one generation frequently changes in the next.

55. There is a great deal of literature on incentive effects. In general, incentive effects improve performance and reduce the effects of bias, although in a few experiments there was no effect or even a negative effect. See Gregory Mitchell, "Why Law and Economics' Perfect Rationality Should Not Be Traded for Behavioral Law and Economics' Equal Incompetence," *Georgetown Law Journal* 91 (2002): 67, 116–17. The predominant weight of such studies thus supports the common-sense view that legislators and experts may update better because of their incentives. Certainly the burden should be on theorists of cultural cognition to show that such incentives do not matter.

56. See, e.g., D. N. Stone and D. A. Ziebart, "A Model of Financial Incentive Effects in Decision Making," *Organizational Behavior and Human Decision Processes* 61 (1995): 250.

57. Mark Seidenfeld, "Cognitive Loafing, Social Conformity and Judicial Review of Agency Rulemaking," *Cornell Law Review* 87 (2001): 486, 516–17. Professor Seidenfeld cites the cognitive literature on this point (516n52).

58. Roger D. Congleton, "Rational Ignorance, Rational Voter Expectations, and Public Policy: A Discrete Informational Foundation for Fiscal Illusion," *Public Choice* 107 (April 2001): 35, 46, http://www.springerlink.com/content/l7508383712758r3/fulltext.pdf.

59. Ray C. Fair, "Econometrics and Presidential Elections," *Journal of Economic Perspectives* 10 (1996): 89–90.

60. Mary Summers and Philip Klinkner, "The Election of John Daniels as Mayor of New Haven," *PS: Political Science and Politics* 23 (1990): 142, 143.

61. Popkin, *Reasoning Voter*, 135.

62. Ibid.

63. Frank Fischer, "Policy Discourse and the Politics of Washington Think Tanks," in *The Argumentative Turn in Policy Analysis and Planning*, ed. Frank Fischer and John Forester (Durham, NC: Duke University Press, 1993), 31.

64. The importance of experts in agenda setting has been on the rise given the complexity of the modern world. Steve Rayner, "Democracy in the Age of Assessment: Reflections on the Roles of Expertise and Democracy in Public-Sector Decision Making," *Science and Public Policy* 30 (2003): 163.

65. Posner, *Law, Pragmatism*, 18.

66. Caplan, *Myth of the Rational Voter*, 50–94.

67. Ibid.

68. Andrew Rich, "The Politics of Expertise in Congress and the News Media," *Social Science Quarterly* 82 (2009): 583.

69. Philip E. Tetlock, *Expert Political Judgment: How Good Is It? How Can We Know?* (Princeton, NJ: Princeton University Press, 2006), 157–58.

70. Stanley Rothman et al., "Politics and Professional Advancement among College Faculty," *Forum* 3 (2005): article 2.

71. *Cf.* John O. McGinnis, et al., "The Patterns and Implications of Political Contributions of Elite Law School Faculty," *Georgetown Law Journal* 93 (2005): 1167, 1995n. 98

72. Allan Meltzer, "Monetary and Other Explanations of the Start of the Great Depression," *Journal of Monetary Economics* 2 (1976): 455, 469.

73. Cf. Ober, *Democracy and Knowledge*, 89.

74. National Commission on Excellence in Education, *A Nation at Risk: The Imperative for Educational Reform*, A Report to the Nation and the Secretary of Education, U.S. Department of Education, April 1983, available at http://teachertenure.procon.org/sourcefiles/a-nation-at-risk-tenure-april-1983 .pdf.

75. J. Johnson, "International Educational Rankings Suggest Reform Can Lift U.S.," U.S. Dept. of Education, Homeroom Blog, Dec. 8, 2010, http:// www.ed.gov/blog/2010/12/international-education-rankings-suggest-reform -can-lift-u-s.

76. For discussion of the common political objectives generated by geopolitical competition, see chapter 2.

77. Patrick McGuinn, "The National Schoolmarm: No Child Left Behind and the New Educational Federalism," *Publius: The Journal of Federalism* 35 (2005): 41–68.

78. Pub. L. No. 107-110 (2001).

79. The No Child Left Behind Act mentions scientific evidence more than a hundred times. See Turner, "Populating a Trials Register," 203, 205.

80. Pub. L. No. 107-279 (2002).

81. Turner, "Populating a Trials Register," 206.

82. Benjamin Michael Superfine, "New Directions in School Funding and Governance: Moving from Politics to Evidence," *Kentucky Law Journal* 98 (2009): 653, 687–88.

83. Dan J. Nichols, "Brown v. Board of Education and The No Child Left Behind Act: Competing Ideologies," *BYU Education and Law Journal* 1 (2005): 151, 175.

84. Michael Podgurksy and Matthew Springer, "Teacher Performance Pay: A Review," *Journal of Policy Analysis and Management* 26 (2007): 909, 931.

85. Ibid.

86. Ibid., 929–31.

87. Ibid., 935–41.

88. U.S. Department of Appropriation Act, Pub.L. No. 109-149 (2006).

89. "Tide Shifting on Merit Pay," *Daily Herald*, Nov. 30, 2010.

90. Podgurksy and Springer, "Teacher Performance Pay," 913–21.

91. William J. Bushaw and Shane Lopez, "A Time for Change: The 42nd Annual Phi Beta Delta/Gallup Survey of Public Attitudes toward Public Schools," *Phi Delta Kappan* 92, no. 1 (2010): 9–26, available at http://www.pdkintl.org /kappan/docs/2010_Poll_Report.pdf.

92. "Charter Schools," National Education Association, http://www.nea.org /home/16332.htm.

93. "Multiple Choice: Charter School Performance in 16 States, Executive Summary," Credo at Stanford University, June 2009, available at http://credo. stanford.edu/reports/MULTIPLE_CHOICE_EXECUTIVE%20SUMMARY.pdf.

94. National Charter School Resource Center, http://www.charterschool center.org/priority-area/understanding-charter-schools. Forty states, along with Washington, D.C., and Puerto Rico, have signed charter school legislation.

95. "Multiple Choice." While the results of this study indicate that there is likely no perfect nation-wide result, independence and variability among charter schools allow for experimentation.

96. Ibid.

97. Lewis-Black et al., *American Voter Revisited*, 405.

## Chapter 9: De-biasing Democracy

1. Jamie Carson et al., "Redistricting and Party Polarization in the U.S. House of Representatives," *American Politics Research* 35 (2007): 878.

2. Bruce Ackerman and Ian Ayres, "The New Paradigm Revisited," *California Law Review* 91 (2003): 743, 763n62.

3. For a discussion of responsiveness, see Adam Cox, "Partisan Fairness and Redistricting Politics," *NYU Law Review* 79 (2004): 751, 765. For an excellent review of the many possible standards of fairness in elections, see Jeanne Fromer, "An Exercise in Line-Drawing: Deriving and Measuring Fairness in Redistricting," *Georgetown Law Journal* 93 (2005): 1547. The focus here is largely on what Professor Fromer describes as competiveness (1583), because competiveness helps promote democratic updating.

4. Bernard Grofman, "Criteria for Districting: A Social Science Perspective," *UCLA Law Review* 33 (1995): 77, 151.

5. Ibid.

6. Ibid.

7. The most important article on this kind of gerrymander is Samuel Issacharoff, "Gerrymandering and Political Cartels," *Harvard Law Review* 116 (2002): 593, 600.

8. Adam Clymer, "The Nation: Democracy in Middle America; Why Iowa Has So Many Hot Seats," *New York Times*, Oct. 27, 2002; Issacharoff, "Gerrymandering," 626.

9. Daryl J. Levinson and Richard H. Pildes, "Separation of Parties, Not Powers," *Harvard Law Review* 119 (2006): 2311, 2381. See also Jamie L. Carson and Michael H. Crespin, "The Effect of State Redistricting Methods on Electoral Competition in United States House of Representatives Races," *State Politics and Policy Quarterly* 4 (2004): 455.

10. Text of Proposition 11, California Redistricting Reform Initiative, codified at California Constitution, Art. XXI, available at http://www.calvoter.org/issues/votereng/redistricting/prop11text.html.

11. Hillel Aron, "Soft Revolution Ends Gerrymandering," *L.A. Times Weekly*, Nov. 4, 2010.

12. See, e.g., "Drawing the Lines," editorial, *L.A. Times*, Sept. 24, 2010, http://articles.latimes.com/2010/sep/24/opinion/la-ed-prop2027-20100924.

13. Jim Tankersley, "Ohio, California Take Divisive Redistricting Wars to Ballot Box," *Toledo Blade*, Oct. 23, 2005, http://www.toledoblade.com/apps/pbcs.dll/article?AID=/20051023/NEWS09/510230348.

14. Corbett A. Grainger, "Redistricting and Polarization: Who Draws the Lines in California?" *Journal of Law and Economics* 53 (2010): 545–67.

15. Ibid., 563.

16. Alex Isenstadt, "California Redistricting Produces New House Hopefuls," *Politico*, June 30, 2011, http://www.politico.com/news/stories/0611/58083.html.

17. Florida Amendment 5 reads in part, "Legislative districts or districting plans may not be drawn to favor or disfavor an incumbent or political party. Districts shall not be drawn to deny racial or language minorities the equal opportunity to participate in the political process and elect representatives of their choice. Districts must be contiguous. Unless otherwise required, districts must be compact, as equal in population as feasible, and where feasible must make use of existing city, county and geographical boundaries." Amendment 6 applied similar language to congressional districts.

18. Richard Pildes, "Why the Center Does Not Hold: The Causes of Hyperpolarized Democracy In America," *California Law Review* 99 (2011): 273.

19. Stephen Macedo, "Toward a More Democratic Congress? Our Imperfect Democratic Constitution: The Critics Examined," *Boston University Law Review* 89 (2009): 609, 620.

20. Nathaniel Persily, "In Defense of Foxes Guarding Henhouses: The Case for Judicial Acquiescence to Incumbent-Protecting Gerrymanders," *Harvard Law Review* 116 (2002): 649, 650.

21. Pildes, "Why the Center Does Not Hold," 298.

22. Elisabeth Gerber and Rebecca Morton, "Primary Election Systems and Representation," *Journal of Law, Economics, and Organization* 14 (1998): 304. See also Barry C. Burden, "Candidate Positioning in U.S. Congressional Elections," *British Journal of Political Science* 34 (Apr. 2004): 211–27.

23. See "Congressional and Presidential Primaries: Open, Closed, Semi-Closed, and 'Top Two,'" Fairvote.org, http://archive.fairvote.org/?page=1801. More than fifteen states have open primaries.

24. Angela Galloway, "State's 'Top Two' Primary Upheld by U.S. Justices," *Seattle Pi*, March 18, 2008, http://www.seattlepi.com/local/article/State-s-Top-Two-primary-upheld-by-U-S-justices-1267609.php.

25. Democratic Party v. Jones, 530 U.S. 567 (2000).

26. Washington State Grange v. Washington State Republican Party, 552 U.S. 442, 445, 450 (2008).

27. Rebecca M. Kysar, "Listening to Congress: Earmark Rules and Statutory Interpretation," *Cornell Law Review* 94 (2009): 519, 534.

28. Thomas E. Mann and Norman J. Ornstein, *The Broken Branch: How Congress Is Failing America and How to Get It Back on Track* (New York: Oxford University Press, 2006).

29. Cf. Adam Feibelman, "Contract, Priority and Odious Debt," *North Carolina Law Review* 85 (2007): 727, 759.

30. Carle Hulse, "Senate Won't Allow Earmarks in Spending Bills," Caucus Blog, *New York Times*, Feb. 1, 2011, http://thecaucus.blogs.nytimes.com/2011/02/01/senate-wont-allow-earmarks-in-spending-bills.

31. Matt Kelley, "Seniority Helps Fund Lawmakers' Pet Projects," *U.S.A. Today*, Feb. 10, 2010, http://www.usatoday.com/news/washington/2010-02-10-earmarks_N.htm.

32. Scott Lilly, "Is the Politics of Pork Poisoning Our Democracy?" *Roll Call*, Aug. 15, 2005, http://www.rollcall.com/issues/51_16/-10294-1 .html?zkMobileView=false.

33. Ibid.

34. Popkin, *Reasoning Voter*, 98.

35. Indeed, Senator Mitch McConnell has argued that earmarks have no effect on the level of spending, just who determines spending. See Jeanne Sahadi, "Earmark Ban: No Effect on Spending," *CNN Money*, Nov. 17, 2010, http:// money.cnn.com/2010/11/15/news/economy/earmarks_ban/index.htm.

36. "Duffy, Obey Trade Barbs on Earmarks," *Chippewa Herald*, Nov. 25, 2010, http://chippewa.com/news/local/article_ccde9f76-f7fb-11df-85ed-001cc4c002e0.html.

37. Moreover, there are constraints on this practice known as "lettermarking." Both Presidents Bush and Obama have issued executive orders instructing agencies not to make grants based on communications with Congress. See Ron Nixon, "Some Earmarks Could Elude a Ban," *New York Times*, Dec. 28, 2010.

38. The strongest argument to date is that term limits would reduce the power of special interests and make legislators more responsive to voters. See Einer El-hauge, "Are Term Limits Undemocratic?" *University of Chicago Law Review* 64 (1997): 83, 116–39.

39. Ibid., 156.

40. Ibid., 86.

41. Quinn Bowman and Chris Amico, "Congress Loses Hundreds of Years of Experience, but Majority of Incumbents Stick Around," PBS Online, Nov. 5, 2010, http://www.pbs.org/newshour/rundown/2010/11/congress-loses-hundreds-of-years-of-experience-but-vast-majority-of-incumbents-stick-around.html.

42. Alexander Tabarrok, "A Survey, Critique, and New Defense of Term Limits," *Cato Journal* 14, no. 2 (1994), available at http://www.cato.org/pubs /journal/cjv14n2-9.html.

43. In the past the older generations may have been able to offer a unique contribution to social deliberation because they simply knew about unrecorded past regularities, like the location of floods. Posner, *Aging and Old Age*, 207–208. But that function has largely disappeared. All information is now recorded, and the old have no special role in storing it. Today, however, the old are more likely to retard democratic updating, because they are more afflicted with confirmation bias than the young, being more reluctant to give up the status quo to which they are accustomed (107).

44. Gerald F. Seib, "In with the New—And Young—Republicans," *Wall Street Journal*, Jan. 6, 2011, http://online.wsj.com/article/SB10001424052748704835504576059582025431402.html#project%3DCONGRESS_AGES_1009.

45. Ibid.

46. As calculated from the ages of newly elected senators found at http:// en.wikipedia.org/wiki/List_of_current_United_States_Senators_by_age.

47. "Legislator Demographics," National Conference of State Legislatures, http://www.ncsl.org/?tabid=14850.

48. U.S. Term Limits v. Thornton, 514 U.S. 779 (1995).

49. While there has been much discussion of term limits for Supreme Court justices (see, e.g., Steven G. Calabresi and James Lindgren, "Term Limits for the Supreme Court: Life Tenure Reconsidered," *Harvard Journal of Law and Public Policy* 29 (2006): 769, 822–75), the problem of accelerating technology is not offered as a rationale.

50. As calculated from their official biographies; see SupremeCourt.gov, http://www.supremecourtus.gov/about/about.html.

51. Steven Smith, "Law without Mind," *Michigan Law Review* 88 (1989): 104, 117.

52. Craig Lerner and Nelson Lund, "Judicial Duty and the Court's Cult of Celebrity," *George Washington Law Review* 78 (2010): 1255, 1271.

53. Martin Lipset, "Some Social Requisites of Democracy: Economic Development and Political Legitimacy," *American Political Science Review* 53 (1959): 69, 80–81. John Stuart Mill saw civic education as essential to broadening the franchise. John Dunn, "Western Political Theory in the Face of the Future" (New York: Cambridge University Press, 1979), 51–53. See also Lydia Saad, "In U.S., 22% Are Hesitant to Support a Mormon in 2012," Gallup Online, June 20, 2011, http://www.gallup.com/poll/148100/Hesitant-Support-Mormon-2012.aspx.

54. Scott O. Lilienfeld, Rachel Ammirati, Kristin Landfield, "Giving Debiasing Away: Can Psychological Research on Correcting Cognitive Errors Promote Human Welfare?" *Perspectives in Psychological Science* 4 (2009): 390, 393.

55. Laura J. Kray and Adam D. Galkinsky, "The Debiasing Effect of Counterfactual Mind-sets: Increasing the Search for Disconfirmatory Information in Group Decisions," *Organizational Behavior and Human Decision Processes* 91 (2003): 69.

56. Currently statistical education in the United States is very weak. Peter Sedlmeier, *Improving Statistical Reasoning: Theoretical Models and Practical Implications* (Mahwah, NJ: Lawrence Erlbaum, 1999), 196. But statistics courses focused on "judgments in everyday life" suggest statistical training could improve statistical reasoning. See Geoffrey T. Fong, David H. Krantz and Richard E. Nisbett, "The Effects of Statistical Training on Thinking About Everyday Problems," in *Rules for Reasoning*, ed. Richard Nisbett (Mahwah, NJ: Lawrence Erlbaum, 1993), 91, 121.

57. Pinker, "Mind Games," 261. Caplan has much the same program. See Caplan, *Myth of the Rational Voter*, 198.

58. Somin, "Voter Ignorance," 419.

59. Popkin, *Reasoning Voter*, 36–37.

60. Druckman, "Politics of Motivation."

61. Robin Hanson, "Idea Futures," *Wired*, Sept. 1995, available at http://hanson.gmu.edu/ideafutures.html.

## Conclusion

1. W. Brian Arthur, *The Nature of Technology: What It Is and How It Evolves* (New York: Free Press, 2009), 200.

2. Morris, *Why the West Rules*, 191.

3. Michael Hanagan and Chris Tilly, eds., *Contention and Trust in Cities and States* (New York: Springer, 2011), 214.

4. John Reader, *Cities* (New York: Grove Press, 2006), 54.

5. Johnson, *Where Good Ideas Come From*, 162.

6. Ober, *Democracy and Knowledge*, 38.

7. Ibid., 70–79.

8. Ibid., 161.

9. Ibid., 142–51.

10. Ibid., 150.

11. Ibid., 219–20.

12. Ithiel De Sola Pool, *Technologies of Freedom* (Cambridge, MA: Belknap Press, 1983).

13. Joel Mokyr, *The Enlightened Economy: An Economic History of Britain, 1700–1850* (New Haven, CT: Yale University Press, 2009), 413–14.

14. Ibid.

15. Voltaire, *Letters Regarding the English Nation* (London: Westminster Press, 1926), 144.

16. Mokyr, *Enlightened Economy*, 418.

17. Peter Temin, "Two Views of the British Industrial Revolution," *Journal of Economic History* 57 (1997): 63, 80.

18. Gordon Wood, *Empire of Liberty: A History of the Early Republic* (New York: Oxford University Press, 2009), 251.

19. Pauline Maier, *Ratification: The People Debate the Constitution, 1787–1788* (New York: Simon and Schuster, 2010), 165.

20. Ibid., ix.

21. Ibid., 59–60.

22. Ibid., 72.

23. Pool, *Technologies of Freedom*, 15.

24. Alexander Hamilton, "Federalist No. 9," *The Federalist Papers* (New York: Mentor, 1991), ed. Clinton Rossiter, 66–71.

25. Ibid.

26. Wood, *Empire of Liberty*, 311.

27. Ibid.

28. Vernor Vinge, a mathematician and science fiction writer, first put forward the idea of the singularity in a speech at a conference sponsored by NASA. "What Is the Singularity?" March 1993, available at http://mindstalk.net/vinge/vinge-sing.html. It has been popularized by Ray Kurzweil, an engineer and inventor. See Kurzweil, *Singularity*.

29. Max More and Ray Kurzweil, "Max More and Ray Kurzweil on the Singularity," Kurzweil Accelerating Intelligence, Feb. 26, 2002, http://www.kurzweilai.net/articles/art0408.html?printable=1.

30. David Gelles, "NASA and Google to Back School for Futurists," *Financial Times*, Feb. 3, 2008, http://www.ft.com/cms/s/0/8b162dfc-f168-11dd-8790-0000779fd2ac.html.

31. The leading theorist of combinatorial prediction markets is Robin Hanson. See, e.g., Robin Hanson, "Combinatorial Information Market Design," *In-*

*formation Systems Frontier* 5 (2003): 107. See also D. M. Pennock and R. Sami, "Computational Aspects of Prediction Markets," in *Algorithmic Game Theory*, ed. Noam Nisan et al. (Cambridge: Cambridge University Press, 2007), 651.

32. See John Holland, "What Is to Come and How to Predict It," in *The Next Fifty Years: Science in the First Half of the Twenty-First Century*, ed. John Brockman (New York: Vintage Books, 2007), 179.

33. Leila Gray, "Gamers Succeed Where Scientists Fail," Futurity, Sept. 29, 2011, http://www.futurity.org/science-technology/gamers-succeed-where-scientists-fail.

# Index

Abramowicz, Michael, 60, 65, 6–69, 70, 71, 112, 161, 176n1, 177n37, 178n44, 189n4
Abrams, Burton A., 32–33
Abrams, Samuel J., 129
Ackerman, Bruce, 139
Adler, Jonathan H., 103, 115
Advisory Commission to Study the Consumer Price Index, 20
agent-based modeling, 107
Aguayo v. Richardson (1973), 56
Alesina, Alexandro, 127, 191n28
Alho, Eeva, 69
Alm, Richard, 19
American Republic, founding of, 154–56
Amico, Chris, 143
Ammirati, Rachel, 146
Anderson, Kenneth, 104, 188n55
Aron, Hillel, 140
Arrison, Sonia, 169n80, 190n20
Arthur, W. Brian, 150, 165n1
artificial intelligence (AI), 9, 94, 119; benefits of in an age of accelerating technology, 105–8; exaggerated fears of, 101–4; Friendly AI, 94, 102, 103, 104, 185n1; the futility of relinquishing AI and prohibiting battlefield robots, 100–101; and mimicking aspects of the human brain, 98; strong AI, 94, 95–99; threats of, 99–100
Athens, 34, 151–52; and the Council of 500, 151–52
Atkin, Charles, 182n45
Atluri, Satya N., 14
Aubuchon, Mark, 73
Aurther, Charles, 54
Axtell, Robert L., 106
Ayres, Ian, 106, 139

Bacon, Francis, 47
Balz, Dan, 130
Bandyopadhyay, Subir, 182n52

Banerjee, Bibek, 182n52
Barberis, Nicholas, 69
Barboza, David, 170n10
Barnes, Robert, 184n69
Barro, Robert, 81–82
Bartels, Larry, 192n34
Bell, Tom W., 73
Bell's law, 10, 12
Bennett, Robert, 78, 126–27
Berg, Joyce E., 71, 176n6
Berners-Lee, Tim, 56–57
bias, 121, 147; biased assimilation, 124; confirmation bias, 124, 129; and cultural cognition and motivated reasoning, 124–25; democracy's capacity to act in the face of bias, 134–36; and experts, 34–35; framing, 125–26, 128; heuristic bias, 69; information technology and bias, 136–37; innate majoritarian bias, 123; "knowledge falsification" by the majority, 122–23; long-shot bias, 6–69; nature of, 121–22; representativeness bias, 69; special interest bias, 122; status quo bias, 123–24, 144. See also constraints on bias
"big data" phenomenon, 105
Bill of Rights, 52–53, 53, 155
Bimber, Bruce, 33, 127
Blais, Lynn E., 114
Bleske, Glen L., 87
Bloch-Elkon, Yael, 129
blogs. See dispersed media
Bollinger, Lee C., 74, 179n64
Bonneau, Chris W., 87
Borgatti, Stephen P., 188n64
Bowman, Quinn, 143
Braman, Donald, 125, 193n50
Brandeis, Louis, 50
Brennan, Geoffrey, 34
Brighton, Henry, 95, 186n14
Britain, 92; and the industrial age, 152–54; and Parliament, 153–54

Brooks, David, 180n14
Brown, Dan J., 135
Brown, Norman R., 121
Brownstein, Ronald, 192n31
Buchanan, Neil H., 189n15, 189–90n17
Buckley v. Valeo (1976), 184n70
Burden, Barry C., 197n22
Burke, Dan, 9
Burke, Edmund, 123
Bush, George W., 130, 134, 198n37; and tax cuts, 27
Bushaw, William J., 135
Butler, Declan, 43
Butler, Henry N., 49

Cain, Bruce, 45
Calabrese, Anthony, 19
Calabresi, Steven G., 198n49
Calculated Risk (website), 77
Cantor, Joseph E., 90
Caplan, Bryan, 123, 133, 193n51
Carney, Ellen Newlin, 90
Carson, Jamie L., 138, 196n9
Cartier, Curtis, 85
catastrophic events, 69, 107, 11–19
Chalmers, David J., 96
charter schools, 54, 135–136
checklists, 188n58
Cherry, Miriam, 73
China, 12, 18, 28
Choi, James J., 46
Citizens United v. FEC (2010), 91, 183n62, 184n70
Civilization (computer game), 158
Clark, Annette, 28
climate change, 3, 114, 115; and prediction markets, 66
Clymer, Adam, 139
Coate, Stephen, 32, 91
Coglianese, Cary, 50
Cohen, Jon, 130
Coleman, John J., 88, 182n44
collective decision making, 36, 37; the "basic public action problem" of, 60; complications introduced by technological innovation, 15; dominance of special interests in, 31–33; and social complexity, 76
Commodity Futures Trading Commission (CFTC), 72, 178n54
computation-driven technological change,

13; energy, 15–16; medicine, 13–14; nanotechnology, 14–15
computer games, 158. See also specific computer games
computer simulations, 106–7, 158
computers/computation, 9; exponential growth in computation, 1, 9–13, 96; and gains in connectivity, 12; and progress in hardware and software, 11. See also computation-driven technological change
"computing capacity of information," 10
Congleton, Roger D., 132
consequentialist democracy, 25–28, 38; and consensus on the general need for collective decision making, 26; and cosmic and moral questions, 84; focus of on evaluating policies, 30; nondemocratic mechanisms as aid to, 3–39; view of gerrymandering, 139
constraints on bias: cognitive diversity, 12–31; democratic institutions, 126–28; experts, 132–33; political representatives, 131–32
Cornell Law School, 45
Cosmides, Leda, 121, 190n3
Courcy, Catherine, 50
Cowen, Tyler, 17, 18, 19, 21
Cox, Adam, 196n3
Cox, W. Michael, 19
Crawford, Susan P., 85
Crespin, Michael H., 196n9
Croley, Stephen P., 187n53
Cromwell, Oliver, 31–32

Daniel, Marcus, 181n32
data: and artificial intelligence (data gathering and hypotheses about data), 105–6; the "big data" phenomenon, 105; and prediction markets, 66–67; transparency and accessibility of, 48, 56–58
Davenport, Thomas H., 54
decentralization, of policy, 2, 40, 48, 49–54, 111–13
Deep Blue (supercomputer), 96
Defense Advanced Research Projects Agency (DARPA), 70, 99
Delbecq, Benoit, 103
Democratic Party v. Jones (2000), 142
Democratic-Republican Party, 156
Dennett, Daniel C., 185n11

Deutsch, David, 5, 78
Diacu, Florin, 107
"digital democracy," 3
DiNardo, J., 42
dispersed media, 1–2, 75, 77–83, 129; and a culture of learning, 83–84; improving factual knowledge of the policy world, 7–79; and social networks, 21, 180n15; specialized blogs by law professors, 179n1
distribution of information, 84–85; avoiding regulations that discriminate against the new media, 84, 85; encouraging universal access, 84, 85–86; getting information to voters, 85, 86–88; providing more money for campaigns, 85, 89–93
Dodd-Frank law, 72
Doe v. Reed (2010), 183n62
Dolinar, Richard, 55
Donahue, John J., 57
Downs, Anthony, 36, 191n18
Drew, Christopher, 105
Druckman, James, 125, 128, 147, 191n17
Dudziak, Mary, 179n7
due process clause, 51
Duffy, John F., 189n2
Dunn, John, 199n53
Dyson, Freeman, 114

economic growth, 27–28; as a valence issue, 27
Economix (blog), 79
Edelstein, Stephen, 97
Edsall, Thomas B., 183n65
education, 26–27, 28; educational initiatives, 4–49 (see also specific initiatives); as a valence issue, 27. See also K-12 education reform
Education Sciences Reform Act (2002), 4–49, 134
Ehrlich, Paul, 66
Eichengreen, Barry, 45
Einhorn, Hillel, 121
elections: and election markets, 67–68; legislative elections, 127; and polls, 35, 67–68; presidential elections, 127. See also political campaigns
Elhauge, Einer R., 115, 143, 198n38
Elms, Lauren, 88

empiricism, 2, 9, 40–43, 83, 84; advantages of, 45; controlling for the possibility that correlations do not reflect causation, 41; declining relative cost of, 33–45; and event studies, 63–64; and field experiments, 42, 54–55; its past and its future, 46–47; limitations on experiments, 42–43; and meta-analyses, 44; and multiple-level modeling techniques, 44; and natural experiments, 41–42; and peer review, 45; and prediction markets, 65–66; and repeated sampling, 44; reporting bias in empirical studies, 57; the rise of empirical analysis, 43–45. See also empiricism, and information-eliciting rules
empiricism, and information-eliciting rules, 4–49, 83–84; changing laws to promote private incentives for research, 58; and the creation of a better political culture, 5–59; rules encouraging decentralization, 49–54; rules encouraging the randomization of the application of different policies, 54–56, 175n69; rules making government data more available in the most transparent and useful form, 56–58
encompassing interests, 31–34; improving incentives to vote on, 34
endowment effect, 124
energy (alternatives to fossil fuels), and computer-driven technological change in, 15–16; solar energy, 15
Epstein, Joshua M., 106
Epstein, Richard A., 49
Erikson, Robert S., 65
eSolar, 15, 167n38
Establishment Clause, 53
Etzioni, Amitai, 183n65
experiments: field experiments, 42, 48, 54–56; limitations on, 42–43; natural experiments, 41–42; psychological experiments, 80
experts, 2, 25–26, 34–36, 46, 60, 66, 75; and constraint of bias, 131–34; and dispersed media, 78; and think tanks, 133

Fair, Ray C., 132, 170n4
Federal Aviation Administration (FAA), 111

Federal Communication Commission (FCC), 54–55, 90
Federal Election Commission (FEC), 89, 181n38
Federal Election Commission v. Wisconsin Right to Life, Inc. (2007), 184n70
federalism, 37, 49–54, 112; advantages of, 49; constitutional federalism, 50; costs of, 49; and "negative externalities," 49; political right's and left's interest in reviving of, 50; reinforcement of its virtues by modern technology, 50. See also decentralization, of policy
Federalism Accountability Act (1999), 173n43
Federalist Papers, 155
Federalist Party, 156
Feibelbaum, Adam, 142
Fermi, Enrico, 159
Fermi paradox, 15–59
Feynman, Richard P., 14
Fienberg, Stephen F., 56
Fiorina, Morris P., 129, 193n43
First Amendment, 32, 51, 85
Fischer, Frank, 132
Fitts, Michael A., 30
Flint, Joe, 90
Foldit (computer game), 158
Fong, Geoffrey T., 199n56
foreign policy events, 70
Fourteenth Amendment, 52
401(k) opt-out plans, 46
framing, 125–26
Franzee, Stephanie, 85
Freedman, David H., 166n13
Freedom of Information Act (1966), 56
Friedel, Robert, 12, 17, 99
Friedland, Steven I., 40
Friedman, Alinda, 121
Friedman, Avi, 20
Friedman, Jeffrey, 67
Fromer, Jeanne, 195–96n3
functionalism, 95
Funk, Carolyn J., 34

Galkinsky, Adam D., 146
Gallagher, Maggie, 52
Galloway, Angela, 141
gambling, 73

Gawande, Atul, 188n58
Geelan, Jeremy, 11
Gelles, David, 157
General Agreement on Tariffs and Trade (GATT), 28
Gentzkow, Matthew, 80
geopolitical competition, 28
Gerber, Alan, 192n32
Gerber, Elisabeth, 142
Gerhardt, Michael J., 79
gerrymandering, 88, 126, 131; gerrymandering reform, 13–41; gerrymandering reform in California, 139–40; gerrymandering reform in Florida, 140; gerrymandering reform in Iowa, 139; two kinds of gerrymandering, 13–39
Gersen, Jacob, 75
Gingrich, Newt, 68
GINI coefficient, 63
Giovannini, Enrico, 30
global warming. See climate change
Go (Chinese game), 96
Godwin, Kenneth R., 183n65
Gomez, Eric, 184n73
Good, Irving John, 95
Google, 54; and the creation of an online library, 58, 175n84; petition to Nevada to make autonomously driving cars legal, 97
government, branches of: executive, 48, 109; judiciary, 48, 109; legislative, 48, 109. See also president, the; U.S. Congress; U.S. Supreme Court
government transparency, 74–75
graduated driver licensing, 46
Grainger, Corbett A., 140
Grasping Reality with the Invisible Hand (website), 77
Gray, Jim, 172n9
Gray, Leila, 158
Green, Donald P., 192n32
Greenberg, David, 54
Grimmelman, James, 125, 193n50
Grofman, Bernard, 139
Gueorguieva, Vassia, 180n15
gun control, 53, 125

Hahn, Robert W., 60, 65
Hall, Melinda Gann, 87
Hall, Ralph, 144

Hamilton, Alexander, 127, 155
Hanagan, Michael, 151
Hand, Learned, 59
Hanson, Robin, 16, 38, 71, 147, 200n31
Hausman, Jerry, 20
Haw, Rebecca, 67, 69
Heald, Gary, 182n45
Held, David, 25
Henderson, M. Todd, 63, 176n14
Hetrick, Carl C., 185n75
Hilbert, Martin, 10
Hogarth, Robin, 121
Holland, John, 158
Holmes, Richard, 99
Homeland Security alerts, 41–42
Hounshell, David, 16
Houser, Daniel, 91
Hsu, Fen-hsiung, 96
Hulse, Carl, 70, 142
Hunter, Dan, 80, 180n20
*The Hurt Locker,* 187n38
Hyman, David A., 58
incentive effects, 194n55

India, 12, 18
information bandwagons, 123
information cascades, 123
information costs, reduction of, 2–29; cre-
    ating common knowledge, 29–30; cre-
    ating incentives to act on encompassing
    interests, 31–34; improving incentives
    to vote on encompassing interests, 34;
    mixing expert and nonexpert opinion,
    34–36
information markets. *See* prediction
    markets
information technology. *See* artificial intel-
    ligence (AI); dispersed media; empiri-
    cism; prediction markets
Institutional Review Boards (IRBs), 58
Intel, 11
Internet, the, 9, 11; access to, 20; data
    available on, 43–44; encouragement
    of dispersed media, 1–2; facilitation of
    prediction markets, 1; as a funnel of in-
    formation, 79–80; and mobile devices,
    20; and online learning, 33; and predic-
    tion markets, 61; and search capacity,
    97–98; the semantic Web, 56–57; struc-
    ture of, 81. *See also* dispersed media
Intrade, 61, 62–63

Iowa Electronic Futures Market, 72,
    176n6
Iran, 101
Iraq War, 129–30
Ireland, 73
Isenstadt, Alex, 140
Islam, failure of to adapt to technological
    disruption, 22–23
Issacharoff, Samuel, 126, 127, 139,
    192n31, 196n7

Jacobson, Gary C., 87, 129, 129–30
Johnson, George, 166n16
Johnson, J., 134
Johnson, Steven, 79, 151
Johnston, David Cay, 19
*Journal of Empirical Legal Studies,* 45,
    172n16
Joy, Bill, 99, 101

K-12 education reform, 134–36, 145–46;
    charter schools, 54, 135–36; merit pay,
    135; school vouchers, 53, 54
Kadak, Andrew C., 116
Kahan, Dan, 125, 193n50
Kahneman, Daniel, 69
Kaku, Michio, 10, 11, 12
Kang, Michael J., 126
Kasparov, Garry, 96, 106
Kaustia, Markku, 69
Kelley, Matt, 142
Kerr, Orin, 179n7
Kierkegaard, Søren, 65
Kinder, Donald R., 128
Klazczynski, Paul A., 130
Kleiner, Keith, 10
Klick, Jonathan, 41–42, 42
Klinkner, Philip, 132
Knauer, Nancy J., 130
"knowledge falsification," 122–23
Koh, W. T. H., 182n52
Krantz, David H., 199n62
Krawitz, David, 20
Kray, Laura J., 146
Krent, Harold J., 55
Krugman, Paul, 81
Kuran, Timur, 123
Kurzweil, Ray, 11, 16–17, 96, 98, 99, 157,
    160, 166n25, 166n28, 189n16
Kysar, Rebecca M., 142

Laibson, David, 46
Lakhoff, George, 125
Landfeld, Kristin, 146
Landis, James M., 109
Langevoort, Donald C., 178n52
Lauerman, John, 13
Law, David S., 107
Lee, Dwight S., 42
Leegin Creative Leather Prods. v. PSKS, Inc. (2007), 46
Leininger, S. Luke, 55
Lemley, Mark, 9
Leonhardt, David, 67
Leppik, David, 114
Lerner, Craig, 145
Leung, H. M., 182n52
Levey, Noam N., 50
Levinson, Daryl J., 139
Levy, Steven, 98
Lewis, Kenneth A., 32–33
Lewis-Beck, Michael S., 26, 124, 136
Lilienfeld, Scott O., 146
Lilly, Scott, 142
Lindblom, Charles, 83
Lindgren, James, 41, 44, 45, 65, 198n49
Linzer, Peter, 51
Lipset, Martin, 145–46
Liptak, Adam, 141
Littleston, Nick, 188n63
Lodge, Milton, 124
Lohmann, Suzanne, 171n21
Lohr, Steve, 98
Lomasky, Loren, 34
Lopez, Edward J., 89
Lopez, Priscila, 10
Lopez, Shane, 135
Lucas, Robert E., Jr., 11
Lund, Nelson, 145, 173–74n47

Macedo, Steven, 140
Macey, Jonathan N., 49
MacLennan, B. J., 166–67n30
Macmillan, Harold, 130
Madison, James, 127
Madrian, Brigette C., 46
Maier, Pauline, 155
Malkiel, Burton G., 178n43
Mandel, Gregory, 15
Mankiw, N. Gregory, 19, 78, 81
Mann, Thomas E., 142
Manna, Paul F., 182n44

Manne, Henry G., 172n16
Manzi, Jim, 54, 55
Marconi, Andrea L., 73
Marcus Aurelius, 45
Marginal Revolution (website), 77
markets: and bubbles, 68, 69–70; information from, 36–37; and long-shot bias, 6–69; market mechanisms and politics, 3. See also prediction markets; stock market, the
Markoff, John, 99
Massachusetts Supreme Court, 52
Masters, William A., 103
Matsusaka, John G., 192n32
McAleer, Michael, 17
McConnell, Mitch, 198n35
McConnell v. FEC (2003), 183n62
McDonald v. Chicago (2010), 52–53, 53
McGuinn, Patrick, 134
Mckenzie, David, 179n5
McQuaid, Brian M., 73
Medicaid, 23
Medicare, 23, 32
medicine, and computer-driven technological change in, 13–14; computers as warning systems, monitors, and maintenance mechanisms, 13; digitization of medical records, 13; sequencing a genome, 13
Mello, Michelle M., 46
Meltzer, Allan, 133
Merritt, Deborah, 53
Metrick, Andrew, 46
Metzger, Gillian E., 50
Meyer, Bruce D., 20
Middlebrook, Stephen, 43
Mihim, Stephen, 179n3
Milbank, Dana, 145
Mill, John Stuart, 199n53
Mitchell, Gregory, 45, 172n18, 194n55
Moe, Terrence, 174
Moe, Terry M., 32
Mokyr, Joel, 12, 47, 153, 165n5
Moore, Gordon, 9
Moore, Max, 157
Moore's law, 9–10, 96, 114; as part of a more general exponential growth in computation, 11; prediction of its imminent death, 11
Moravec, Hans, 11
More, Max, 157
Morgan, Gareth, 17

Morris, Ian, 4, 16, 150
Morris, Stephen, 32
Morton, Rebecca, 142
Movsesian, Mark L., 28, 70
Moynihan, Patrick, 38
Muggleton, Stephen H., 105
Muller, Jerry Z., 170n16
Muller, John, 43
multiverse politics, 156–59

Naam, Ramez, 15
"naïve realism," 125
nanotechnology, and computer-driven
    technological change in, 14–15; 3-D
    printing, 14–15; machines that will
    self-assemble, 14
Naone, Erica, 57
A Nation at Risk (National Commission
    on Excellence in Education), 134
National Charter School Resource Center,
    136
National Commission on Excellence in
    Education, 134
Nelson, Kjersten R., 191n17
Nelson, Phillip, 32
Nelson, Thomas E., 192n35
Neteller, 73
New Deal, 2, 110
New York State federal court of appeals,
    56
Newell, Allen, 95
Newman, Alexandra Lee, 70, 71, 178n46
newspapers, 82; in early America, 154;
    partisan control of, 82
Nichols, Dan J., 135
nirvana fallacy, 67, 70, 71, 112
Nisbett, Richard E., 199n56
Niv, Y., 122, 190n4
Nixon, Ron, 198n37
No Child Left Behind Act (2001), 48, 134,
    173n32, 195n79
nonexperts, 34–36, 66
North Korea, 101
Northwest Austin Municipal Utility Dis-
    trict No. 1 v. Holder (2009), 141
Noveck, Beth Simone, 56

Obama, Barack, 3, 49, 57, 62, 198n37;
    campaign against al-Qaeda, 100; and
    government transparency, 75, 179n66;

health care reform of, 50, 64; and the
    killing of Osama bin Laden, 68; Open
    Government Initiative of, 57; State
    of the Union address (January 2010),
    184n69; stimulus plan of, 27
Ober, Josiah, 27, 34, 60, 134, 151–52
Office of Information and Regulatory Af-
    fairs (OIRA), 109–10
Office of Minority Health, 19
Olson, Mancur, 170n17, 191n18
Omohundro, Steve, 103
Oprea, Ryan, 71
O'Reilly, Charles A., III, 121
originalism, online debate over, 78,
    179nn6–7
Ornstein, Norman J., 142
Ott, Michael, 61
Ozler, Berk, 179n5

PACER, 57
parliamentary government, 152–53
Partridge, Derek, 44
Pascal's wager, 119
Paypal, 73
peer review, 45, 103, 187n53
Pericles, 25
Perry, William J., 101
Persico, Nicola, 183n59
Persily, Nathaniel, 140
Pickert, Kate, 19
Pildes, Richard H., 127, 139, 140, 141
Pinker, Steven, 125, 146
Podgurksy, Michael, 135
political campaigns: focus of on policy
    consequences, 87–88; the Lincoln-
    Douglas debates, 86; political cam-
    paign advertising, 85, 86–88; political
    campaign spending, 86–87; providing
    more money for political campaigns,
    85, 89–93
political parties, as coalitions, 131
political representatives, and constraint of
    bias, 131–34
Pool, Itheil Sola de, 31, 153, 155
Pope, Alexander, 30
Popkin, Samuel L., 66, 67, 79, 80, 132,
    142, 146, 171n28, 177n29
Porat, Avner M., 121
Posner, Richard A., 25–26, 86, 92, 114,
    118, 119–20, 124, 132, 177n23,
    178n48, 190n21, 198n43

Predator (robotic drone), 99, 100, 104
prediction markets, 1, 35, 38, 60, 81, 83,
    84, 114, 117, 119–20; advantages of,
    65–67; conditional event markets,
    61–63; how they temper three of the
    largest problems of politics, 60; and the
    improvement of regulatory decisions,
    112–13; legalization and subsidiza-
    tion of, 72–74; and multiverse politics,
    157–58; nature of, 60–64; and the
    problem of teasing correlation apart
    from causation, 63–64; provision of
    conditions for through government
    transparency, 74–75; and reduction in
    ideological bias, 146–48; as a response
    to social complexity, 76; responses to
    critiques of, 67–72; the use of subsidies
    to encourage experiments in predica-
    tion market design, 74
president, the: and executive orders, 110;
    and "lettermarking," 198n37; role of in
    creating information-eliciting rules, 51
President's Council of Advisors on Science
    and Technology, 11
Price, Vincent, 182n53
primary voting reform, 141–42; open pri-
    maries, 141; top two primaries, 141
Protestantism, 32
public goods, 26. See also education
Puttonen, Vesa, 69
Pythagoras, 43

Race to the Top Initiative, 49, 136
Rachlinski, Jeffrey J., 129, 192n41
radical life extension, 22, 119, 190n22
Raines, S. Travis, 69
randomization, of policy, 40, 48, 54–56
Rappaport, Michael B., 174n56, 179n6,
    191n21
Rayner, Steve, 194n64
Reader, John, 151
Reagan, Ronald, 110; "Morning in
    America" campaign, 87
regulation, in an age of technological
    acceleration, 2, 109–11; accounting
    for possible catastrophes and Elysian
    benefits, 11–20; decentralizing central-
    ized regulation, 111–13; and intergen-
    erational equity, 116–18; the present
    future of, 113–16
Rehnquist, William, 49

Reinberg, Steven, 19
religions: effects of accelerating technol-
    ogy on, 21 (see also Islam, failure of to
    adapt to technological disruption); and
    globalization, 22
rent-seeking, 31
representative democracy, 28, 131–32, 152
research, promoting private incentives for,
    40, 58
retail price maintenance, 46
Reynolds, Glenn, 79
Rich, Andrew, 133
Ridout, Travis N., 182n44
Rietz, Thomas, 71, 176n6
Roberts, John, 49
Robinson, Bill, 130
Rogers, Robert L., 73
Romney, Mitt, 68
Rose, Lowell C., 28
Rosen, Sherwin, 19
Rosenberg, David, 92
Rosenthal, Howard, 127, 191n28
Roth v. United States (1957), 52–53
Rothman, Stanley, 133
Rowbottom, Jacob, 92
Rucker, Randall R., 171n21
Rumsfeld, Donald, 130


Saad, Lydia, 199n53
Sahadi, Jeanne, 198n35
Sahuguet, Nicolas, 183n59
Saletan, William, 193n53
Samaha, Adam M., 175n71
same-sex marriage, 52; and generational
    updating, 130, 193n52
Samuelson, Pamela, 175n84
Sanandaji, Tino, 81
Sanders, Lynn M., 128
Savage, Charles, 181n37
Scarborough, Ryan, 100
Schoen, Douglas E., 68
Schudson, Michael, 34
Schwartz, John, 57
scientific method, 3
Searle, John R., 95–96
Sedlmeier, Peter, 199n56
Seeley, Katharine Q., 75
Seib, Gerald F., 144
Seidenfeld, Mark, 131, 194n57
Selina, Howard, 95, 186n14
Shachtman, Noah, 99

Shapiro, Ben, 184n75
Shapiro, Jesse M., 80
Shapiro, Robert Y., 129
Sharkey, Catherine, 173n44
Shays v. FEC (2004), 181n38
Shier, Maria Tess, 17
Shiller, Robert, 24
SimCity (computer game), 158
Simon, Herbert A., 66, 95
Simon, John, 44
Simon, Julian, 66
Singer, P. W., 100
singularity, 156–57, 200n28; technological singularity, 157
Singularity Institute for Artificial Intelligence, 185n1, 187n54
Singularity University, 157
Skokowski, Paula, 17
Slade, Margaret E., 46
Slotjee, Daniel, 17
Smith, Adam, 153, 170n16
Smith, Bradley, 181n38, 181–82n41
Smith, David, 145
Smith, John R., 68
Smith, Steven, 145
Smoot-Hawley Tariff, 133
Sniderman, Paul M., 88
Snowberg, Erik, 64, 69
Sobel, Russell S., 69
social knowledge, 2, 5, 25, 29–30, 40; comparison to the knowledge of natural science, 5; and the consequentialist function, 29–30; dispersed media's structuring of for social use, 79–80; improvement of, 23–24; increase of through prediction markets, 65–66; and issues amenable to factual resolution, 28; and the preference-eliciting function, 29–30; as provisional, 5
social networks. See dispersed media
Social Security, 23, 32, 116–17
Solum, Lawrence B., 185n3
Somin, Ilya, 100, 128, 146, 192n33
special interests, 31–33, 71, 90, 110, 127; and the creation of purely regulatory solutions, 115; effect of technological change on, 33; effect of term limits on, 144; the elderly, 32; oil companies, 32; the power of special interest funding, 91; special interest bias, 122; teachers unions, 32
Springer, Matthew, 135

Srivastava, Deepak, 14
Stamp Act (1765), 31–32
Stanovich, Keith, 192n42
Stephenson, Neal, 12
Stern, Philip M., 183n65
Steuerle, C. Eugene, 23
Steuerle-Roper index of fiscal democracy, 23
Stevik, Kristin, 88
stock market, the, 69–70
Stone, D. N., 194n56
Stratmann, Thomas, 86, 90, 91, 182n43, 182n47, 183n65
Strauss, Aaron B., 128
Sullivan, Andrew, 174n48
Sullivan, James X., 20
Summers, Larry, 19–20, 174n62
Summers, Mary, 132
Sunstein, Cass, 65, 80–81
Superfine, Benjamin Michael, 49, 134
Surowiecki, James, 35
Swert, Knut de, 86, 182n42
Szlay, Alexander, 172n9

Tabarrok, Alexander, 41–42, 42, 144
Taber, Charles, 124
Taft-Hartley Act (1947), 184n70
Taleb, Nassim Nicholas, 108
Tankersley, Jim, 140
Taub, Stephen, 46
technological acceleration, 3, 9, 40; advantages of, 35; the case for, 16–21; and decentralization of policy, 52; and dispersed media, 83; future technological change and current regulation, 113–16; and "the law of accelerating returns," 16–17; and measures of economic growth in the United States, 1–20; and the need for prioritization of policies, 36; and social complexity, 76; and a technological "singularity," 156–57
technological disruption, 4; social remedies for the problems of, 21–24
technological domain, 165n1
technological innovation, 3; and innovations in social governance, 149–51. See also American Republic, founding of; Athens; Britain: and the industrial age
technological singularity, 157

technology, 3; stage 1 technology, 10; stage 2 technology, 10; stage 3 technology, 10; stage 4 technology, 10; "technologies of freedom," 31, 33
Tedeschi, E., 82
television, 80, 82; partisan media on cable television, 82
Temin, Peter, 153
Tetlock, Philip E., 78, 133
Thiel, Peter, 17
Thompson, William N., 73
Thurman, Walter N., 171n21
Tillman Act (1907), 184n70
Tilly, Chris, 151
Tocqueville, Alexis de, 154
Tooby, John, 121, 190n3
Topol, Eric, 13, 15
Turner, Henry A., 185n75
Turner, Herbert, 49, 134, 173n32, 195n79
Tversky, Amos, 69
*2001: A Space Odyssey*, 95

Unarmed Aerial Vehicles (UAVs), 100. *See also* Predator (robotic drone)
Unlawful Internet Gambling Enforcement Act (UIGEA) (2006), 73
updating, 121–22, 129, 193n51, 198n43; aspects of cognition that improve updating of information, 131; constraint on by the status quo bias, 124; the diversity of citizens in democratic updating, 12–31; generational updating, 130–31, 193n54; in policy innovation, 82; in technological innovation, 82
U.S. Congress, 2; age among members of, 144–45; and the delegation of power, 109–10; and earmarks, 142–43; elimination of funding for the Open Government Initiative, 57; and government transparency, 75; and limitations on political campaign contributions, 89–90; role of in creating information-eliciting rules, 4–49, 51; and term limits, 143–45
U.S. Constitution, 154–55; constitutional amendment process, 53–54. *See also* Bill of Rights; First Amendment; Fourteenth Amendment

U.S. Supreme Court, 2; and the case for term limits, 145–46; and due process jurisprudence, 51–52; and limitations on political campaign contributions, 89–90; the Rehnquist Court, 53; on retail price maintenance, 46; the Roberts Court, 53; role of in creating information-eliciting rules (a jurisprudence of social discovery), 51–53; ruling on term limits, 145; rulings on primaries, 142. *See also specific Court cases*
U.S. Term Limits v. Thornton (1995), 145

valence issues, 27
value issues, 27
Van Aelst, Peter, 86, 182n42
Verne, Jules, 185n4
Vezina, Kenrick, 10
Vietnam War draft lottery, 56
Vinge, Vernor, 200n28
Voltaire, 153
von der Fehr, Nils-Henrik M., 88
von Hayek, Friedrich, 65
voting: and the "Cook Partisan Voting Index," 197n23; the diversity of voters in democratic updating, 12–31; extreme voters, 127; getting information to voters, 86–88; improving incentives to vote, 34; median voters, 126–28; and runoff, 87; swing voters, 128; weak partisan voters, 127–28. *See also* political campaigns; primary voting reform

Wade, Nicholas, 16
Wagner, Wendy E., 114
*Wall-E*, 95
War Labor Disputes Act (1943), 184n70
Warmuth, Manfred, 188n63
Warsh, David, 5
Warwick, Kevin, 186n31
Washington State Grange v. Washington State Republican Party (2008), 142
Watson (supercomputer), 4, 97
Wayner, Peter, 43
websites. *See* dispersed media
Weisenthal, Joe, 81
Weitzman, Martin, 12
West, Richard F., 192n42
Whitney, Lance, 20

Wikileaks (website), 181n37
Wlezien, Christopher, 65
Wolf, Patrick J., 174n54
Wolfe, Alan, 181n33, 193n43
Wolfers, Justin, 57, 61, 63, 69, 176n5,
    176n7, 176–77n17
Wood, Gordon, 156
worldviews, 124, 130, 193n54
Wright, Joshua, 172n16
Wright, Robert, 102

X Prize Foundation, 116
xenophobia, 123

Young, Iris Marion, 29
Young, Jeffrey R., 80

Zaller, John, 182–83n53
Zarefsky, David, 86
Zeiler, Kathryn, 46
Zelman v. Simmons-Harris (2002), 53
Zeppos, Nicholas S., 55
Zhao, Xinshu, 87
Ziebart, D. A., 194n56
Zitzewitz, Eric W., 61, 63, 176n5, 176n7,
    176–77n17
Zlauvinen, Gustavo R., 101